Computer and Communication Systems Performance Modelling

C.A.R. Hoare, Series Editor

BACKHOUSE, R.C., *Program Construction and Verification*
BACKHOUSE, R.C., *Syntax of Programming Languages: Theory and practice*
DEBAKKER, J.W., *Mathematical Theory of Program Correctness*
BARR, M. and WELLS, C., *Category Theory for Computing Science*
BEN-ARI, M., *Principles of Concurrent Programming*
BIRD, R. and WADLER, P., *Introduction to Functional Programming*
BJÖRNER, D. and JONES, C.B., *Formal Specification and Software Development*
BORNAT, R., *Programming from First Principles*
BUSTARD, D., ELDER, J. and WELSH, J., *Concurrent Program Structures*
CLARK, K.L. and McCABE, F.G., *micro-Prolog: Programming in logic*
CROOKES, D., *Introduction to Programming in Prolog*
DROMEY, R.G., *How to Solve it by Computer*
DUNCAN, F., *Microprocessor Programming and Software Development*
ELDER, J., *Construction of Data Processing Software*
ELLIOTT, R.J. and HOARE, C.A.R., (eds.) *Scientific Applications of Multiprocessors*
GOLDSCHLAGER, L. and LISTER, A., *Computer Science: A modern introduction (2nd edn)*
GORDON, M.J.C., *Programming Language Theory and its Implementation*
HAYES, I. (ed). *Specification Case Studies*
HEHNER, E.C.R., *The Logic of Programming*
HENDERSON, P., *Functional Programming: Application and implementation*
HOARE, C.A.R., *Communicating Sequential Processes*
HOARE, C.A.R., and JONES, C.B. (ed), *Essays in Computing Science*
HOARE, C.A.R., and SHEPHERDSON, J.C. (eds). *Mathematical Logic and Programming
   Languages*
HUGHES, J.G., *Database Technology: A software engineering approach*
INMOS LTD. *occam 2 Reference Manual*
JACKSON, M.A., *System Development*
JOHNSTON. H., *Learning to Program*
JONES, C.B., *Systematic Software Development using VDM (2nd edn)*
JONES, C.B. and SHAW, R.C.F. (eds), *Case Studies in Systematic Software Development*
JONES, G., *Programming in occam*
JONES, G. and GOLDSMITH, M., *Programming in occam 2*
JOSEPH, M., PRASAD, V.R. and NATARAJAN, N., *Multiprocessor Operating System*
LEW, A., *Computer Science: A mathematical introduction*
MacCALLUM, I., *UCSD Pascal for the IBM PC*
MARTIN, J.J., *Data Types and Data Structures*
MEYER, B., *Introduction to the Theory of Programming Languages*
MEYER, B., *Object-oriented Software Construction*
MILNER, R., *Communication and Concurrency*
MORGAN, C., *Programming from Specification*
PEYTON JONES, S.L., *The Implementation of Functional Programming Languages*
POMBERGER, G., *Software Engineering and Modula-2*
REYNOLDS, J.C., *The Craft of Programming*
RYDEHEARD, D.E. and BURSTALL, R.M., *Computational Category Theory*
SLOMAN, M. and KRAMER, J., *Distributed Systems and Computer Networks*
SPIVEY, J.M., *The Z Notation: A reference manual*
TENNENT, R.D., *Principles of Programming Languages*
WATT, D.A., *Programming Language Concepts and Paradigms*
WATT, D.A., WICHMANN, B.A. and FINDLAY, W., *ADA: Language and methodology*
WELSH, J. and ELDER, J., *Introduction to Modula-2*
WELSH, J. and ELDER, J., *Introduction to Pascal (3rd edn)*
WELSH, J., ELDER, J. and BUSTARD, D., *Sequential Program Structures*
WELSH, J. and HAY, A., *A Model Implementation of Standard Pascal*
WELSH, J. and McKEAG, M., *Structured System Programming*
WIKSTRÖM, Å., *Functional Programming using Standard ML*

Computer and Communication Systems Performance Modelling

Peter J. B. King

Prentice Hall
New York   London   Toronto   Sydney   Tokyo   Singapore

First published 1990 by
Prentice Hall International (UK) Ltd
66 Wood Lane End, Hemel Hempstead
Hertfordshire HP2 4RG
A division of
Simon & Schuster International Group

Printed and bound in Great Britain
at the University Press, Cambridge.

Library of Congress Cataloging-in-Publication Data

King, Peter J.B., 1952—
    Computer and communication systems performance
    modelling/Peter J.B. King.
    p. cm.
    ISBN 0-13-162984-0: $67.95
    1. Electronic digital computers — Evaluation.
    2. Telecommuncation systems — Evaluation. I. Title.
    QA76.9.E94K56   1989   89-28401
    004.2-dc20   CIP

British Library Cataloguing in Publication Data

King, Peter J.B., 1952—
    Computer and communcation systems performance
    modelling.
    1. Computer systems. Networks, Queues. I. Title
   004.6

ISBN 0-13-162984-0
ISBN 0-13-163065-2 pbk

2 3 4 5   94 93 92 91

For Rosemary

Contents

Preface

Outline of book

This book is an introductory text in the field of quantitative analysis of computer and communication system performance. Knowledge of elementary probability theory and simple stochastic processes is used, but the material is also briefly outlined in the early chapters. Although no knowledge of queueing theory is assumed, a number of aspects which only receive passing mention in other books will make this a useful reference book for the professional as well as a student text.

The first chapter introduces Kendall's (extended) notation for describing queueing systems, and sample path arguments that lead to Little's result, which relates queue lengths to arrival rates and waiting times. Various queueing disciplines are discussed qualitatively. The elements of probability theory are reviewed in the second chapter. This should be an element of revision for the student, since it is by no means a full introduction to the theory but covers all that is needed for this book. Chapter 3 gives a brief introduction to the theory of stochastic processes, in particular Markov processes, which play a central role in the analyses presented in later chapters. Little's theorem is proved more formally.

The body of the book starts in chapter 4, with simple queues. $M/M/1$ queues are examined in detail, starting with the time dependent formulation, and then leading on to the steady state solution. Generalisations such as the machine-repairman problem and the finite source queue are considered. Simple discrete time queues are also considered. The next chapter treats queues in which the service times can be generally distributed, and three different approaches are used to derive the Pollaczek-Khintchine formula. The following chapter treats systems which are unreliable, and have servers that break down and can be repaired. This study involves non-trivial manipulations of the generating functions and is perhaps the simplest system in which such calculations arise naturally. Priority queues, in which some jobs have priority over others form the subject of chapter 7. Both preemptive and non-preemptive priorities are considered. Kleinrock's conservation law, which provides a measure of the cost of different priority assignments, is proved. The next chapter uses the Laplace transform to evaluate busy period distributions. These non-trivial manipulations of Laplace transforms also allow waiting time distributions to be found for several different scheduling disciplines.

Chapter 9 considers multiserver Markovian queues. The analysis of $M/M/1$ queues is extended to $M/M/c$ queues, and taken to the limit as a study of $M/M/\infty$ queues. A number of two server systems with unusual characteristics are analysed.

Systems in which the two servers have different speeds are studied, and also systems in which the second server becomes available only after a delay. Multiserver priority systems are analysed.

Chapter 10 looks at networks of queues. The effect of connecting the output of one queue as the input to another is investigated. The ideas of local balance and time reversibility allow a large class of networks to be analysed. Solutions to networks in this class have the property known as *product form*, since the steady state probability of a network state is the product of the steady state probabilities of the nodes in the network considered in isolation. The following chapter examines some of the elegant algorithms for computing various performance metrics for networks of the product form class. We examine some of these and discuss their limitations. Chapter 12 treats approximation techniques based on product form solutions, and derives a number of bounds on throughput and delay which can be calculated more cheaply than full solutions.

Numerical methods for the solution of queueing problems form the subject of the next chapter. These include Gaussian elimination on the balance equations, eigen-analysis of the transition matrix, and Courtois type decomposition methods. Finally, a chapter on local area network performance is included.

Most chapters contain exercises which may aid understanding. Each chapter has a bibliography, giving the sources for the techniques described, and suggestions for further reading. I have endeavoured to reference only widely available journals and books, rather than technical reports, even when these are in fact the original publication.

The book is designed to be suitable for final year undergraduates in computer sciences, or conversion course MSc students. It will also find use in electrical engineering courses on communication systems analysis, and in mathematics departments as a source of examples in the field of operations research.

Prerequisites for the study of this book are an introductory course in probability theory, such as is given to most science and engineering students. An understanding of both discrete and continuous distributions is required, along with the elementary properties of generating functions, distribution and density functions, Laplace transforms, and moments. The necessary material is briefly covered in chapters 2 and 3.

Acknowledgments

This book would not have been possible without the support and encouragement of many people. I have been privileged to work closely with Isi Mitrani for many years, and he taught me most of what I know in this subject area. He will recognise some of our joint work appearing here in book form. Nachum Shacham will also see his influence on many pages. The Department of Computer Science at Heriot-Watt University provided a working environment that encouraged me to write the book and allowed the class testing of some portions of the material. The computer facilities were also provided by Heriot-Watt University. The major portion of the book was typed by the author, although first drafts of chapters 5 and 8 were typed by Mrs Shirley Storer. Leslie Lamport's LaTeX system was used to typeset the book, with the final camera-ready copy being produced at the University of London Computer Centre on the Linotronic L-300 phototypesetter. Helen Martin from Prentice-Hall attempted to keep me on schedule and did not despair when deadline after deadline was passed.

The book was read at various stages by Ilze Ziediņš, Stan Zachary, Peter Thanisch, Rob Pooley, Jane Hillston, and Saqer Abdel-Rahim. They clarified many points, and corrected errors. Their suggestions, and those of the publisher's anonymous reviewers, have been invaluable. Despite their efforts, errors undoubtedly remain, for which the responsibility rests with the author.

Finally, I must thank my wife, Rosemary, and children, Robert and Anna, for putting up with many husbandless and fatherless evenings and weekends when I was writing. Without their tolerance and encouragement this book would not have been finished.

Edinburgh   PJBK

Chapter 1

Introduction

Computer and communication systems are expensive resources, and it is important that the most efficient use is made of them. Although the necessity to keep the CPU active as much as possible is perhaps less important today, the increasing use of sophisticated communications systems has led to many demands to share scarce resources. It is important to be able to quantify the performance of systems both existing and at the design stage, so that cost/benefit type of analyses can be used to choose between design alternatives.

This book is an introduction to the performance modelling of computer and communication systems. We shall be examining such questions as:

1. How much buffer space should be provided for users' input?
2. Will adding a second CPU to the system be more effective than upgrading the current CPU to a higher speed one?
3. What response time can we expect to simple queries on the database?
4. How many processes will be waiting for execution on average?
5. What retransmission policy will optimise throughput in a CSMA local area network?

There are three approaches to this type of problem. First, the system can be built, or modified, and then measured. If the measurement is performed while a standard set of tasks is running, this technique is known as *benchmarking*. This has the advantage that performance estimate is perfectly accurate. No errors have been introduced that are not there in the real system. The disadvantages of benchmarking are the cost and inflexibility of the technique. It will be expensive to acquire new equipment just for the purposes of evaluation. Even if it can be borrowed, there will be expense involved in configuring software to use the new equipment. Each benchmark will usually run for a fairly long period in order that its measurements are statistically reliable. If there are a large number of design alternatives, it will be very time consuming to investigate them all.

The second approach is to build a simulation model of the system. This model can be validated against the existing system and then altered to reflect the proposed modifications. In common with benchmarking, simulation is an experimental science; parameters are changed and then the benchmark is performed again or the simulation model is rerun and the performance measured. Simulation models have the property that arbitrary levels of detail can be included. This is both an advantage and a disadvantage. The advantage is that any system feature can be included, to see if that feature has an impact on system performance. The major disadvantage is that there is a great temptation to include many features of the real system in the simulation model, which while adding to realism, also add to the running time of the model and hence the cost of the performance study.

This book deals only with the third approach to performance evaluation and prediction. A mathematical model of the system, or of parts of it, is constructed. This model can then be validated in the same way as a simulation model. After validation, changes to the system can be evaluated by changing appropriate model parameters and reevaluating the model solution. The expense here is in the time of the skilled modeller to construct a model which can both be solved and capture the significant behaviour of the system.

Although computers are completely deterministic machines, our analyses will use the theory of probability extensively. In fact, computer performance modelling has been one of the driving forces behind many recent advances in applied probability theory, particularly in the field of queueing theory. There are two reasons why we use probability theory so extensively. First, many of the processes involved are inherently random. For example, the time that a user takes to type a line of input will vary dramatically depending on their skill as a typist, the complexity of response expected, and many other factors. Similarly, the successful transmission of a packet in a computer network is a random event since there is a chance that there will be some interference with the communication, and it will need to be retransmitted. Secondly, it is more convenient. Even if we had complete knowledge of all the times involved at each stage of each process, the size of the data needed as input to the model would be extremely large, and, perhaps more importantly, the results and conclusions that we could draw would be needlessly specific. By representing the essential characteristics of the system with a few parameters we can evaluate the performance and hope to demonstrate the effects of changes in the parameter values.

1.1   History

The application of probability theory to these problems goes back to 1917 and the Danish engineer A.K. Erlang, who analysed the behaviour of simple queues to assist him in designing telephone exchanges. This telephonic flavour continued until the mid 1960s when the first analysis of a computer time-sharing system was undertaken by Scherr.

Many of the problems that we shall investigate are approached using some aspects of the theory of queues. This theory has been developed over the last seventy years, and extensively so in the last twenty years. The literature is vast, with comprehensive bibliographies having over nine hundred entries in the early 1960s. We shall only touch on some of the simpler aspects of the theory. Over the last twenty years the field has blossomed, and now has several journals devoted exclusively to research papers on the subject.

1.2 Performance measures

There are many reasons why we may wish to calculate the performance of our systems, but for the most part long run costs and revenues are the interests of management. Freak occurrences that happen only in peculiar circumstances are of little interest. We will be examining the long run performance of systems. The notion of *statistical equilibrium* or *steady state* will often be used. This does not necessarily mean that the system behaves in a particularly 'steady' manner, but that the probabilities of the system being in a particular state have settled down and are not changing with time.

When a user submits a job to a system, the question that usually needs to be answered is 'How long before I have my answers?' Customer waiting time and its distribution is one of our main interests. Sometimes, particularly when priority queues are under consideration, we shall want the waiting time conditioned on the service requirement of the customer. The management of the system will be interested in the number of jobs waiting – since they must be stored somewhere – the throughput of the system, how many jobs it can process in unit time, indicates the rate at which revenue will be generated if a charge is made to each job using the system. The utilisation of the system, the proportion of the time that the system is idle, gives some indication of the scope for increasing throughput and hence revenue, without increasing the resources committed. Note that an increase in throughput may only be possible at the expense of increased delay to customers. The distribution of the lengths of idle and busy periods may also be of interest to the management.

1.3 Notation

A queueing system can be described in terms of six component parts.

1. The arrival process: a stochastic process describing how jobs arrive in the system from the outside world.
2. The service process: a stochastic process describing the length of time that a server will be occupied by a job.
3. The number of servers and their rates of service.

4. The queue discipline: the rules for deciding which job or jobs to serve. Sometimes this will only be applicable when the server finishes serving its current
   job, at other times the current job in service may be interrupted and a new job
   take over the server.
5. The waiting room: whether or not there is a limit to the number of waiting jobs
   that can be accommodated.
6. The customer population: this may be considered as part of the arrival process,
   but the number of potential customers may change the rate of arrivals.

A widely used notation for describing queueing systems is Kendall's notation, which
consists of a series of letters and numbers separated into fields by /. The generic
Kendall notation is:

$$A/B/c/n/p$$

The first letter, A, and the second letter, B, describe the arrival and service processes,
respectively. The letters are chosen from a small set of descriptors. The number of
servers is given by c. The final two fields are optional, and if omitted are assumed
to be ∞. If the fourth field, n, is specified, it gives the size of the waiting room.
Conventionally, the waiting room includes the customers at the servers. The final
position is used to indicate the population of potential customers. If there is only a
small number of customers, then the presence of a large proportion of them in the
system will reduce the arrival rate to the system. Conversely, if the population is so
large that the number of customers waiting does not affect the rate at which new
customers arrive, then the potential population is set to ∞.

The letters which are used to describe the arrival and service processes are
usually chosen from the following set:

D standing for deterministic, or constant interarrival times or service times.
M designating that the time intervals are Markovian. In a continuous system, ser-
     vice times are drawn from an exponential distribution. If the system is being
     modelled in discrete time, then service intervals are geometrically distributed.
     Arrivals form a Poisson stream.
G meaning generally distributed. No restrictions are placed on the type of distri-
     bution.

The simplest queueing system that we deal with is the $M/M/1$ system, that is a
single server giving exponentially distributed service times to customers that arrive
in a Poisson stream. A more complex example is $M/D/c/n$ where there are c servers
giving a constant service time to arrivals from a Poisson stream, but the waiting room
is limited to n positions, including those in service. The system studied by Erlang
in 1917 was an $M/M/c/c$ system in which any customer arriving and finding all

c servers occupied was lost. $M/G/1$ specifies Poisson arrivals to a single server, which gives arbitrarily distributed service to each customer.

The notation as presented above is the accepted version of Kendall's notation. Notice that no mention is made of the parameters of the service or arrival processes, nor is any queueing discipline specified. As with all notation, a compromise between concise description and an all encompassing specification has been struck. The notation is sometimes extended in an ad hoc manner by authors. A common addition is a superscript on the arrival or service process descriptors to indicate bulk arrivals or service.

The literature is less clear about its definitions of *waiting time* and *queueing time*. In this book, waiting time is used to denote the time from a job entering a queueing system until it leaves, including the time it spends receiving service. Some authors call this queueing time, or response time. In this book, queueing time will represent that time which is spent in the system before any service is received.

1.4 *D/D/*1 **queue**

The simplest possible 'queueing' system that can be imagined is the $D/D/1$ queue. Arrivals to the server occur at regularly spaced intervals. Each arrival has an identical service requirement. Suppose that the time between successive arrivals is a, and the service requirement of each arrival is s. Both a and s are non-negative quantities. The number of jobs in the system is entirely determined by a, s, and the number of jobs initially present. Consider the case of an initially empty system, and the first arrival occurring at time $t = 0$. If $s < a$ the server will be occupied with the job for s and then idle until time a. It is then busy until $a + s$ and idle from $a + s$ until $2a$. This pattern of alternating busy and idle periods continues for ever. There is never more than a single job present in the system, and if present it is being served. The utilisation of the server is s/a. If $s > a$ then the server will be busy until s with the initial job. The second arrival, at time a, was to wait until s to start service. The queue length is two for a period $s - a$. If $s < 2a$ then the queue length will drop to one until $2a$ at which time it will be two again. It is easily seen that the server will never be idle. The queue length will grow without bound. Clearly, it is a deterministic function of time. A little reflection shows that the number in the system, $N(t)$, satisfies:

$$N(t) = \begin{cases} 0 & s < a \text{ and } na + s < t \le (n+1)a \text{ for some } n \\ 1 & s < a \text{ and } na < t \le na + s \\ 1 & s = a \\ \lfloor t/a \rfloor - \lfloor t/s \rfloor & s > a \end{cases} \tag{1.1}$$

If $s = a$ the server will never be idle, but the queue will never grow either; each arrival occurs at exactly the instant the previous job departs.

Table 1.1 Arrivals during sample period

Job	Arrival Time	Service Time
1	3	6
2	7	4
3	10	8
4	16	5
5	35	3
6	42	7
7	43	4
8	60	5
9	67	4
10	70	4

If there is an initial queue at $t = 0$, then the queue will certainly never empty if $s \geq a$. If $s < a$ then the queue will become smaller. After the server has been idle for the first time, the behaviour described above will take place.

The $D/D/1$ queue is only of interest in that it demonstrates the obvious phenomenon that the arrival rate of work to the system must be less than the rate at which the system can serve it, if the system is not to have an unbounded queue of unfinished work.

1.5   Little's theorem

Some properties of queueing systems are extremely general and depend on very weak assumptions about the behaviour of the various processes. The most important of these results is known as Little's theorem after the author of the first formal proof. It was well known to early workers in queueing and inventory theory but not proved in any formal sense until 1961. It relates the average number of customers in a system to the average time they spend in the system. We examine the number of customers in a queue for a particular example system with given arrival times and service requirements. For simplicity we assume integral arrival and service times. A sample of length 80 seconds is considered, during which only 10 jobs were observed. Their arrival times and service requirements are given in Table 1.1. Examining the whole sample, there are 10 arrivals in 80 seconds, so the average rate of arrivals is 1 every 8 seconds. We denote this by λ. If we assume that jobs are served in the

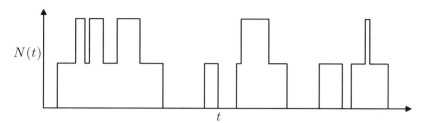

Figure 1.1 Queue length in a FCFS queue

order in which they arrive, so-called first-come-first-served (FCFS) discipline, then the changes in state of the system are given in Table 1.2. Figure 1.1 shows the queue length as a function of time. Examining the number of customers in the system, both in the queue and at the server, we find that there are none present for a total of 30 seconds, there is only a single customer for 33 seconds, and that there are two customers for 17 seconds. The average number is 67/80. Adding the residence times of the customers, we find that it totals 67 seconds. The average value of a customer's residence time is thus 6.7 seconds. The average number in the system, N, satisfies the relationship $N = \lambda W$. Observe that if we make the observation period shorter, the relationship still holds. It holds even if the observations do not start or finish during an idle period, so long as the number of arrivals includes the number present when observations start, and the residence time includes only the time within the observation period.

If we consider the same set of customers arriving at a system, at the same times, but have a different scheduling discipline for the server, the values of N and W may be different, but the relationship will hold. Table 1.3 shows the state changes of the system when the same set of customers arrive and the scheduling discipline is 'last-come-first-serve-preemptive-resume' (LCFSPR). Figure 1.2 shows how the queue length varies with time in this case. That is, a new arrival will be served immediately, with any customer currently in service being forced to wait. The customer which is displaced from the server will resume its service at the point of interruption when all customers which have arrived after it have completed service. Examining the number of customers present in the system, we find that the server was idle for 33 seconds, as before. There was a single customer present for 25 seconds, two for 12 seconds, three for 8 seconds, and four customers in the system for 5 seconds. The average is 93/80. The average residence time was 9.3 seconds. Again we have the relationship $N = \lambda W$.

This relationship between N, the average number in the system, λ, the average arrival rate in the system, and W, the average residence time in the system, holds in *any* system in which these averages exist. It is instructive to verify it for a system

Table 1.2 Changes in state for FCFS queueing

Time	Event and comments
0	Start of sample period
3	Job 1 arrives and starts service
7	Job 2 arrives and waits
9	Job 1 ends service, job 2 starts service
10	Job 3 arrives and waits
13	Job 2 ends service, job 3 starts service
16	Job 4 arrives and waits
21	Job 3 ends service, job 4 starts service
26	Job 4 ends service, server idle
35	Job 5 arrives and starts service
38	Job 5 ends service, server idle
42	Job 6 arrives and starts service
43	Job 7 arrives and waits
49	Job 6 ends service, job 7 starts service
53	Job 7 ends service, server idle
60	Job 8 arrives and starts service
65	Job 8 ends service, server idle
67	Job 9 arrives and starts service
70	Job 10 arrives and waits
71	Job 9 ends service, job 10 starts service
75	Job 10 ends service, server idle
80	End of sample period

which has 'last-come-first-served-preemptive-restart' discipline, which is the same as LCFSPR, except that jobs which are interrupted must resume their service at the beginning. Clearly this will give longer residence times. A heuristic justification of the result can be given when an FCFS system is considered. Consider an 'average' customer. It will spend the 'average' time in the system, W, and leave behind the 'average' number of customers, N. But these customers are just those that arrived during W at the rate of λ per unit time, hence $N = \lambda W$.

It is appropriate here to observe that the proportion of the time that the processor is idle is independent of queueing discipline, so long as the discipline is *conservative*, that is, no work performed by the server is lost, and the server is not idle when there is work to be done. A discipline which does not have this property is

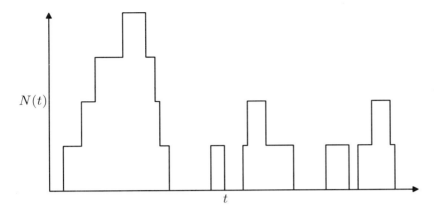

Figure 1.2 Queue length in a LCFSPR queue

'last-come-first-served-preemptive-restart', in which an interrupted job has to start its service from the beginning again. The *busy periods* are exactly the same provided the queueing discipline is conservative. We can see that under the LCFSPR discipline the average waiting time is longer than it is with FCFS in our sample. This may not be true in general, but we can expect that FCFS will have less variance in its waiting times than LCFS.

1.5.1   Proof of Little's theorem

The original proof of Little's theorem is, to say the least, opaque. For a number of years a cottage industry developed in 'simpler proofs of Little's theorem'. The proof we give below is based heavily on our intuition about how numbers in a system and times in a system evolve. A more formal proof is given in section 3.5.

Consider the evolution of a system in time. Customers arrive at intervals and later depart. Notice that no mention need be made of any distribution of time intervals. Little's theorem applies to any particular sample of a system, as well as to the long term averages. Let $A(t)$ denote the number of arrivals to the system before time t. Similarly, let $D(t)$ represent the number of departures from the system before t. Figure 1.3 shows an example of the $A(t)$ and $D(t)$ functions for some system. Clearly, the number of customers in the system at time t, $N(t)$, is given by:

$$N(t) = A(t) - D(t)$$

The total amount of time spent in the system, measured in customer-seconds, is:

$$W(t) = \int_0^t N(s)\,\mathrm{d}s$$

Table 1.3 Changes in state for LCFSPR queueing

Time	Event and comments
0	Start of sample period
3	Job 1 arrives and starts service
7	Job 2 arrives and starts service, job 1 waits
10	Job 3 arrives and starts service, job 2 waits
16	Job 4 arrives and starts service, job 3 waits
21	Job 4 ends service, job 3 resumes
23	Job 3 ends service, job 2 resumes
24	Job 2 ends service, job 1 resumes
26	Job 1 ends service, server idle
35	Job 5 arrives and starts service
38	Job 5 ends service, server idle
42	Job 6 arrives and starts service
43	Job 7 arrives and starts service, job 6 waits
47	Job 7 ends service, job 6 resumes
53	Job 6 ends service, server idle
60	Job 8 arrives and starts service
65	Job 8 ends service, server idle
67	Job 9 arrives and starts service
70	Job 10 arrives and starts service, job 9 waits
74	Job 10 ends service, job 9 resumes
75	Job 9 ends service, server idle
80	End of sample period

Now the number of arrivals is $A(t)$, so the mean time that a customer spends in the system before t is $W_t = W(t)/A(t)$. The mean value of the number in the system, before t is:

$$N_t = \frac{1}{t} \cdot \int_0^t N(s)\, \mathrm{d}s$$

The mean arrival rate before t is $\lambda_t = A(t)/t$. It follows that:

$$N_t = \lambda_t W_t$$

Now W_t will tend to the customer mean residence time, so long as the system empties infinitely often. Taking the limits as $t \to \infty$, Little's result follows:

$$N = \lambda W$$

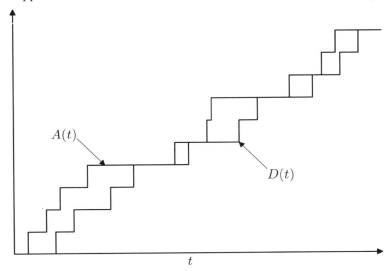

Figure 1.3 Sample of $A(t)$ and $D(t)$

1.6 Applications of Little's theorem

Little's result can be used in many ways to aid in the analysis of computer and communication system performance. Notice that the theorem does not define the 'system'. So long as we are consistent in our definitions and measurements, we can define the system boundary in the most convenient place for the problem at hand. Two examples of the power of Little's theorem are given. The first example is an analysis of the $G/G/1$ queue, the single server queue with arbitrarily distributed arrival and service intervals. The analysis gives a simple relationship between arrival and service rates and server utilisation. The second example uses Little's theorem to derive the response time of a multiuser computer system as a function of the number of users and their mean service requirements.

1.6.1 $G/G/1$ queue

Consider the $G/G/1$ system. We assume that customers arrive at rate λ and have average service requirement $\bar{S}$. We make *no* assumptions about the distribution of service times or interarrival times. If the system is to reach a steady state, then the average amount of work brought into the system in unit time, $\lambda\bar{S}$, must be strictly less than the amount of service that the system can provide in unit time, 1. We call $\lambda\bar{S}$ the traffic intensity, and usually denote it by ρ.

Let us consider the probability that the server is idle, and denote it by π_0. We consider two 'systems'; the complete system of queue and customer in service

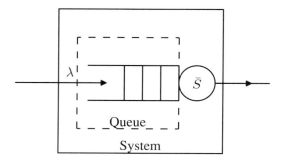

Figure 1.4 $G/G/1$ system

(if any); and the queue alone without the customer being served. The two 'systems' are shown in Figure 1.4. Denote the mean number in the complete system by N, and in the queue alone by Q. Analogously denote the mean residence times in the system by W and the mean time spent in the queue by V. What relationships hold between these quantities? First consider the system residence time, and the queueing time. For *any* customer in the system, customer i say, the difference between the system residence time, w_i, and the queueing time, v_i, is the customer's service requirement, $s_i = w_i - v_i$. Since that relationship holds for each customer, it must also hold for the means of the quantities involved, so $\bar{S} = W - V$. Now consider the numbers of customers in both systems. If there are any customers present, then the number in the system will be one larger than the number in the queue; if there are no customers in the system, then there are none in the queue either. So we can conclude that $N - Q = 1 - \pi_0$. However applying Little's result to the complete system gives $N = \lambda W$. Applying it to the queue only, the mean arrival rate is the same, λ, but the mean queueing time is V, so $Q = \lambda V$. Hence, $N - Q = \lambda(W - V) = \lambda \bar{S}$, and we can conclude that for *any* single server queueing system, $\pi_0 = 1 - \rho$. Notice that this result is independent of queueing discipline. Like Little's result itself, it applies to any sample of a queueing system and can be observed in our earlier sample data. Although the FCFS and LCFSPR systems had different mean queue lengths and mean waiting times, the *busy periods*, that is intervals when the server was active, were identical in both systems, as were the periods when the systems were empty. This is a reflection of the fact that both disciplines were conservative and did not abandon work which had already been performed. The LCFS preemptive-restart discipline is non-conservative, since interrupted jobs lose the work already performed on them, and start at the beginning again. The busy periods will be of different duration.

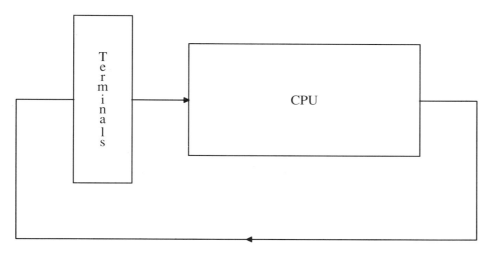

Figure 1.5 Time sharing system

1.6.2 Time sharing system response time

Little's theorem can also be applied to the analysis of the average response time experienced by users of a time sharing system. Consider the system shown in Figure 1.5. There are a fixed number of terminals, N, which submit requests to the central computer system. Once a terminal has submitted a request, it must wait until it receives the response to that request before submitting another. There is a single CPU, which serves requests at the same speed, until there are no requests left to serve. After a request has been served, the terminal user thinks for an interval and then submits another request. The response time to requests is the time that a request spends at the central computer. Without specifying the distribution, assume that the average 'think time', that is time from the response to the previous request until the submission of the next request, at each terminal is $1/\gamma$. Each request has a mean length of $1/\mu$ instructions, and the computer can process c instructions per second. How can the mean response time to a request be found? Let the mean response time be T, and apply Little's theorem to the 'system' consisting of the central computer. Suppose that there are, on average, n requests at the central computer. On average there are $N - n$ terminals thinking, so the rate at which requests are submitted to the computer is $(N - n)\gamma$. Hence:

$$n = (N - n)\gamma T \tag{1.2}$$

This has expressed one unknown, T, in terms of another, n. Now apply Little's theorem to the terminals. The average number in the 'system' is $N - n$, and the average time spent there is $1/\gamma$. The rate at which requests return to the terminals is

$c\mu$, so long as there is at least one request at the central computer system. Denote by π_0, the probability that all requests are at the terminals, hence:

$$N - n = \frac{(1 - \pi_0)c\mu}{\gamma} \tag{1.3}$$

Combining equations (1.2) and (1.3), the following formula for the mean response time, T, is found:

$$T = \frac{N}{(1 - \pi_0)c\mu} - \frac{1}{\gamma} \tag{1.4}$$

This formula is independent of the form taken by the distributions, it depends only on the means. The unknown parameter π_0 will be affected by the distributions of the think times and request lengths. This expression allows the response time of a system to be bounded in a very simple manner. The only parameter that can be altered, by a sophisticated scheduling algorithm say, is π_0. From the point of view of the terminal users, the response time will be reduced if the probability that a request arriving at the CPU finds it idle, which is π_0, is reduced. The minimum possible value of *any* probability is zero, so it follows that:

$$T \geq \frac{N}{c\mu} - \frac{1}{\gamma}$$

Clearly, the response time cannot be less than the service time of a request, so:

$$T \geq \max(\frac{N}{c\mu} - \frac{1}{\gamma}, \frac{1}{c\mu})$$

1.7   Further reading

An overall view of computer performance measurement, modelling and improvement can be had from Svobodova[44] or Borovits and Neumann[3]. Hellerman and Conroy[15] is a slightly older book covering the same material. The techniques of measuring a computer system's performance are covered in Drummond[9] and McKerrow[27]. Simulation modelling is covered by many textbooks. A straightforward introduction, drawing many of its examples from computer systems performance, is given by Mitrani[28]. MacDougal has written specifically on techniques for simulating computer system performance[25]. Fishman's books[11, 12] are excellent introductions to the statistical analysis of simulation results.

As mentioned earlier, the field of performance modelling has grown explosively in the last twenty years, and this is reflected in the literature. Scherr's pioneering thesis was an early publication[38]. Books on queueing theory in general, such as Cox and Smith[8], Saaty[36], Riordan[35], Takács[45], and Cohen[6] were the only available references until the mid 1970s. About this time a number of books appeared applying the techniques of applied probability to computer systems. Coffman and Denning[5], Allen[1], and Kobyashi[22] are all good books from this period. Perhaps the most important book in the field of queueing theory, as applied to computing and communications problems, is the two volume set by Kleinrock[20, 21]. In the 1980s a number of more introductory books have been published such as those by Trivedi[47], Mitrani[29], Leung[23], and Molloy[30]. Schwartz's books[39, 40] are primarily aimed at telecommunications applications, as is the more pragmatic work of Stuck and Arthurs[43]. Takagi has compiled a collection of state of the art papers[46]. The bibliography includes books on queueing theory[7, 31, 14, 33, 34, 48], operations research[16, 17], and more advanced aspects of applied probability and stochastic processes[4, 32, 2, 49]. Sauer and Chandy[37] and Gelenbe and Mitrani[13] are typical of more advanced books on computer system modelling.

Little's theorem was well known for many years, but was first proved by Little in 1961[24]. Simpler proofs were later published by Jewell[19] and Eilon[10]. Maxwell[26] investigated the conditions under which the theorem will hold. Stidham generalised the result to a number of similar situations[41, 18] and also published a proof as the 'last word' on the theorem[42].

1.8 Bibliography

[1] A.O. Allen. *Probability, Statistics, and Queueing Theory: With Computer Science Applications*. Academic Press, New York, NY, 1978.

[2] S. Asmussen. *Applied Probability and Queues*. Wiley, New York, NY, 1987.

[3] I. Borovits and S. Neumann. *Computer Systems Performance Evaluation*. Lexington Books, Lexington, Massachusetts, 1979.

[4] M.L. Chaudhry and J.G.C. Templeton. *A First Course in Bulk Queues*. Wiley, New York, NY, 1983.

[5] E.G. Coffman and P.J. Denning. *Operating Systems Theory*. Prentice-Hall, Englewood Cliffs, NJ, 1973.

[6] J.W. Cohen. *The Single Server Queue*. North Holland, Amsterdam, second edition, 1980.

[7] R.B. Cooper. *Introduction to Queueing Theory*. North-Holland, Amsterdam, second edition, 1981.

[8] D.R. Cox and W.L. Smith. *Queues*. Methuen, London, 1961.

[9] M.E. Drummond. *Evaluation and Measurement Techniques for Digital Computer Systems*. Prentice-Hall, Englewood Cliffs, NJ, 1973.

[10] S. Eilon. A simpler proof of $L = \lambda W$. *Operations Research*, 17(5):915–916, September/October 1969.

[11] G.S. Fishman. *Concepts and Methods in Discrete Event Digital Simulation*. Wiley, New York, NY, 1973.

[12] G.S. Fishman. *Principles of Discrete Event Simulation*. Wiley, New York, NY, 1978.

[13] E. Gelenbe and I. Mitrani. *Analysis and Synthesis of Computer Systems*. Academic Press, London, 1980.

[14] D. Gross and C.M. Harris. *Fundamentals of Queueing Theory*. Wiley, New York, NY, 1974.

[15] H. Hellerman and T.E. Conroy. *Computer System Performance*. McGraw-Hill, New York, NY, 1975.

[16] D.P. Heyman and M.J. Sobel. *Stochastic Models in Operations Research Volume I: Stochastic Processes and Operating Characteristics*. McGraw-Hill, New York, NY, 1982.

[17] D.P. Heyman and M.J. Sobel. *Stochastic Models in Operations Research Volume II: Stochastic Optimization*. McGraw-Hill, New York, NY, 1984.

[18] D.P. Heyman and S. Stidham. The relation between customer and time averages in queues. *Operations Research*, 28(4):983–994, July/August 1980.

[19] W.S. Jewell. A simple proof of $L = \lambda W$. *Operations Research*, 15(6):1109–1116, November/December 1967.

[20] L. Kleinrock. *Queueing Systems Volume 1: Theory*. Wiley, New York, NY, 1975.

[21] L. Kleinrock. *Queueing Systems Volume 2: Computer Applications*. Wiley, New York, NY, 1976.

[22] H. Kobayashi. *Modeling and Analysis*. Addison-Wesley, Reading, Massachusetts, 1978.

[23] C.H.C. Leung. *Quantitative Analysis of Computer Systems*. Wiley, Chichester, 1988.

[24] J.D.C. Little. A proof of the queueing formula $L = \lambda W$. *Operations Research*, 9(3):383–387, May/June 1961.

[25] M.H. MacDougall. *Simulating Computer Systems: Tools and Techniques*. MIT Press, Cambridge, Massachusetts, 1987.

[26] W.L. Maxwell. On the generality of the equation $L = \lambda W$. *Operations Research*, 18(1):172–174, January/February 1970.

[27] P. McKerrow. *Performance Measurement of Computer Systems*. Addison-Wesley, Reading, Massachusetts, 1987.

[28] I. Mitrani. *Simulation Techniques for Discrete Event Systems*, volume 14 of *Cambridge Computer Science Texts*. Cambridge University Press, Cambridge, 1982.

[29] I. Mitrani. *Modelling of Computer and Communication Systems*, volume 24 of *Cambridge Computer Science Texts*. Cambridge University Press, Cambridge, 1987.

[30] M.K. Molloy. *Fundamentals of Performance Modeling*. Macmillan, New York, NY, 1989.

[31] P.M. Morse. *Queues, Inventories and Maintenance*. Wiley, New York, NY, 1958.

[32] G.F. Newell. *Applications of Queueing Theory*. Chapman and Hall, London, 1971.

[33] N.U. Prabhu. *Queues and Inventories*. Wiley, New York, NY, 1965.

[34] N.U. Prabhu. *Stochastic Storage Processes*. Springer-Verlag, Berlin, 1980.

[35] J. Riordan. *Stochastic Service Systems*. Wiley, New York, NY, 1962.

[36] T.L. Saaty. *Elements of Queueing Theory*. McGraw-Hill, New York, NY, 1961.

[37] C.H. Sauer and K.M. Chandy. *Computer System Performance Modeling*. Prentice-Hall, Englewood Cliffs, NJ, 1981.

[38] A.L. Scherr. *An Analysis of Time-Shared Computer Systems*. MIT Press, Cambridge, Massachusetts, 1967.

[39] M. Schwartz. *Computer-Communication Network Design and Analysis*. Prentice-Hall, Englewood Cliffs, NJ, 1977.

[40] M. Schwartz. *Telecommunication Networks: Protocols, Modeling and Analysis*. Addison-Wesley, Reading, Massachusetts, 1987.

[41] S. Stidham. $L = \lambda W$: A discounted analogue and a new proof. *Operations Research*, 20(6):1115–1126, November/December 1972.

[42] S. Stidham. A last word on $L = \lambda W$. *Operations Research*, 22(2):417–421, March/April 1974.

[43] B.W. Stuck and E. Arthurs. *A Computer and Communications Network Performance Analysis Primer*. Prentice-Hall, Englewood Cliffs, NJ, 1985.

[44] L. Svobodova. *Computer Performance Measurement and Evaluation Methods: Analysis and Application*. American Elsevier, New York, NY, 1976.

[45] L. Takács. *Introduction to the Theory of Queues*. Oxford University Press, Oxford, 1962.

[46] H. Takagi, editor. *Stochastic Analysis of Computer and Communication Systems*. North Holland, Amsterdam, 1990.

[47] K.S. Trivedi. *Probability and Statistics with Reliability, Queueing and Computer Science Applications*. Prentice-Hall, Englewood Cliffs, NJ, 1982.

[48] J.A. White, J.W. Schmidt, and G.K. Bennet. *Analysis of Queueing Systems*. Academic Press, London, 1975.

[49] R.W. Wolff. *Stochastic Modeling and the Theory of Queues*. Prentice-Hall, Englewood Cliffs, NJ, 1989.

Chapter 2

Probability theory

In this chapter we review some elementary concepts of probability theory, which are used throughout the rest of the book.

Probability theory deals with the analysis of occurrences which are unpredictable to a greater or lesser extent. These random events, although individually unpredictable, display a great degree of regularity and predictability when examined in large numbers. For example, although it is impossible to predict the outcome of a single spin of the roulette wheel, or the result a single roll of a pair of dice, when a large number of such random events are considered, regular patterns can be expected to emerge. If no 'six' has appeared after 50 rolls of a pair of dice, one may with justification, begin to doubt the fairness of the dice. Probability theory allows exact numerical statements to be made about the likelihood of particular occurrences, or sequences of occurrences. Whether the unpredictability of the outcomes arises through some inherent randomness, or because of our inability to measure or specify the conditions under which the event occurs with sufficient accuracy, is a philosophical point which will not be pursued here. Probability theory can be developed by axiomatising the results that might be expected when conducting experiments on idealised 'fair' dice or tossing 'fair' coins.

2.1 Axioms of probability

The mathematical basis for probability has three important components:-

1. A *sample space*, S, which represents all possible outcomes of the experiment.
2. A collection of *events*, E, which are sets of points in the sample space.
3. A mapping, P, from events to the real numbers. This mapping, P, is a probability mapping if it satisfies the following three conditions:

    (a) For all events A, $A \in E$, the mapping P is defined, and satisfies
    $$0 \le P(A) \le 1$$

    (b) $P(S) = 1$.

    (c) If A and B are mutually exclusive, that is they contain no sample points in common, then $P(A \cup B) = P(A) + P(B)$.

It may help to give a simple example. Throwing a single dart at a board has a certain finite set of outcomes. The dart can fail to score; it can land in the segment corresponding to 1, 2, 3, . . . , 20, or within the corresponding double or treble ring; or it can score a single or a double bullseye. Each of these points in the sample space has a probability, which will depend to a greater or lesser extent on where the dart is aimed, and on the skill of the player. However, events which may be of interest are 'score less than 10' or 'score any double'. Since the possible outcomes we described are mutually exclusive, the probability of *compound events*, such as 'score less than 10', is easily calculated as the sum of the probabilities of the outcomes which lead to that condition. In this case,

$$
\begin{aligned}
P(\text{`Score less than 10'}) = \ & P(1) + P(2) + \cdots + P(9) \\
& + P(\text{`Double 1'}) + P(\text{`Double 2'}) \\
& + P(\text{`Double 3'}) + P(\text{`Double 4'}) \\
& + P(\text{`Treble 1'}) + P(\text{`Treble 2'}) \\
& + P(\text{`Treble 3'}) + P(\text{`Miss board!'})
\end{aligned}
$$

From the probability axioms and set algebra it is possible to derive various other properties of probabilities. We shall merely state them.

The probability of the union of two events, A and B, which are *not* mutually exclusive is given by:

$$P(A \cup B) = P(A) + P(B) - P(A \cap B).$$

The probability of the complement of event A, denoted by $\neg A$, is:

$$P(\neg A) = 1 - P(A)$$

An event which cannot occur has probability 0.

When given a set of events, in default of any other information, we usually assign the points of the sample space equal probability. For example, when tossing a coin, we usually assume that heads and tails are equiprobable, and assign each a probability of $\frac{1}{2}$. Similarly, when rolling a (cubic) die, we assume that each face is equiprobable and has probability $\frac{1}{6}$ of landing uppermost. Probabilities are assigned to compound events by counting the number of possible elementary outcomes which correspond to the compound event. After shuffling a pack of playing cards, the number of possible orderings of the pack is 52!, since any of the 52 cards can be first, then there are 51 possibilities for the next card, 50 for the next, and so on. Hence, the probability of any particular order is the reciprocal of that, 1/(52!). The events of interest are usually subsets of the sample space. For example, consider the probability that the top card in the well shuffled pack is the Ace of Spades. The number of orderings of the well shuffled pack that have the Ace of Spades on top is 51!. The probability of this event is just 51!/(52!), which is 1/52 as one might expect. Notice that the probability of *any* particular card, say the 3 of Clubs, is exactly the same. It is important to distinguish between sampling *with* replacement, and sampling *without* replacement. When rolling a dice, we are taking samples with replacement. The act of taking a sample does not change the probability distribution of the following sample. If cards are removed from the pack after sampling, then the probabilities change. Initially, the probability of the top card being a club, say, is 1/13. After a number of samples have been taken, then the probability will depend on how many of those samples were themselves clubs. In order to implement sampling with replacement in the pack of cards, the card that was turned over must be returned to the pack at a random position.

The analysis of card games, which was the original motivation for the study of probability theory, provides many examples. Consider the game of bridge. Cards are dealt from the top of the pack to each of four players in turn, until all the pack has been used. What is the probability that each player has a full suit, consisting of all thirteen cards of the same suit? One's initial idea might be that there is only one ordering that gives this, but a little thought shows that the order in which the cards are given to players is immaterial. The first card can be of any suit, the second must be a different suit, so there is a choice of 39 cards, the third must be of one of the other suits, a choice of 26 cards, and the fourth must be a particular suit, so there is a choice of 13 cards. Hence, there are $52 \cdot 39 \cdot 26 \cdot 13$ orderings that give the first four cards of different suits. Thereafter, we have no choice as to which suit the cards belong to, so there are only 12 choices for the fifth, sixth, seventh and eighth cards. The probability that a deal at bridge will result in each player having a complete set of one suit, is given by $\frac{52 \cdot 39 \cdot 26 \cdot 13 \cdot 12^4 \cdot 11^4 \cdots 2^4 \cdot 1^4}{52!}$.

2.1.1 Conditional probability

If we know that some event has occurred, then this affects the probability that other events have occurred. For example, if we know that the dart has landed on the top half of the dart board, the probability that it actually scored 1 is greater than if we have no knowledge of its position on the board. The *conditional probability* of A occurring given that B has occurred is

$$P(A|B) = \frac{P(A \cap B)}{P(B)} \tag{2.1}$$

Obviously if A and B are mutually exclusive $P(A|B)$ is 0. If B must have occurred in order for A to have occurred, then $P(A \cap B) = P(A)$ and $P(A|B)$ is $P(A)/P(B)$. Two events are *independent* if knowledge of the occurrence of one of them tells us nothing about the probability of the other; that is:

$$P(A|B) = P(A)$$

An equivalent way of expressing the independence of two events is:

$$P(A \cap B) = P(A) \cdot P(B)$$

A set of events, $A_1, A_2, \ldots, A_n$, is said to be mutually independent if every subset of the events, $A_{i_1}, A_{i_2}, \ldots, A_{i_k}$, satisfies:

$$P(\bigcap_{j=1}^{k} A_{i_j}) = \prod_{j=1}^{k} P(A_{i_j})$$

Note that one can have $P(A \cap B \cap C) = P(A) \cdot P(B) \cdot P(C)$, without $P(A \cap B) = P(A) \cdot P(B)$, $P(B) \cap P(C) = P(B) \cdot P(C)$, or $P(A \cap C) = P(A) \cdot P(C)$. It is also possible for the events to be independent when considered two at a time, and yet not be mutually independent.

2.1.2 Bayes theorem

It is often the case that one can calculate the probability of an event of interest, A say, conditioned on one of the events B_i having occurred. If the complete set of B_i exhausts the sample space, or at least exhausts that subset of it in which A can occur, then:

$$P(A) = \sum_{i} P(A|B_i)P(B_i)$$

This is known as the theorem of *total probability*.

The definition of conditional probability given in equation (2.1) can be used to demonstrate Bayes theorem:

$$P(B_j|A) = \frac{P(A|B_j) \cdot P(B_j)}{P(A)} = \frac{P(A|B_j) \cdot P(B_j)}{\sum_i P(A|B_i)P(B_i)}$$

This theorem is used to calculate the probabilities of a particular event occurring after some information has been discovered about events that have occurred by experiment.

Consider a computer centre in which it is known that the accounts department submits a quarter of the jobs to the system, and the engineering department the rest of the jobs. Hence, the probabilities that a job has been submitted by the engineering department is 0.75, and by the accounts department is 0.25. The probability that an engineering job uses FORTRAN is 0.9, and that an accounting job uses FORTRAN is 0.05. The currently active job is using FORTRAN. What is the probability that it comes from the engineering department? Intuitively, the high proportion of engineering jobs using FORTRAN, and the fact that the majority of jobs come from the engineering department anyway leads one to expect that it is very likely to be an engineering job, but Bayes theorem allows this to be quantified. Denote the events, 'engineering job' by E, 'accounting job' by A, and 'uses FORTRAN' by F.

$$\begin{aligned} P(E|F) &= \frac{P(F|E) \cdot P(E)}{P(F|E) \cdot P(E) + P(F|A) \cdot P(A)} \\ &= \frac{0.9 \times 0.75}{0.9 \times 0.75 + 0.05 \times 0.25} \\ &= 0.675/0.6875 = 0.98181 \end{aligned}$$

2.2   Random variables

A *random variable* is a mapping from the sample space to the real numbers. For example, when the experiment is rolling a die, the sample points are which face is uppermost, and a natural random variable is to map the event 'face 3 uppermost' to the real value 3. In other situations we may wish to associate the random variable with the gain or loss that an event will lead to or the length of time that an event will last.

Usually we associate probabilities with random variables in the natural manner. That is, the probability that a random variable, $\Xi(\omega)$, is equal to r is just the probability of the union of all events that lead to $\Xi(\omega) = r$. This is a perfectly reasonable definition for random variables which take discrete values, such as the natural numbers, but for random variables which take values from a continuous set, the probability of being exactly equal to some value is 0. We shall deal with discrete random variables first, and then extend our treatment to continuous random variables.

2.3   Discrete random variables

Discrete random variables take on values which are usually from the natural numbers, although any finite or countable set of values will do. We usually discuss them in terms of a probability distribution called the *induced* probability distribution. We talk about 'the probability that x equals 3' meaning the probability of the set of events that have the random variable x equal to 3.

Although the probability of each set of events may be needed, it is often sufficient to calculate a number of values which can be derived from the probability distribution. The *expectation* of a random variable, ξ, which has value x_i with probability p_i is given by:

$$E[\xi] = \sum_{\forall i} x_i \cdot p_i$$

Loosely, it might be termed the average value that the random variable will have. Further information can be given in terms of the *moments* of the distribution. The nth moment (nth *crude moment*) is given by:

$$M_n[\xi] = \sum_{\forall i} x_i^n \cdot p_i$$

The first moment of a distribution is the expectation of the distribution. Other classes of moments can also be defined, of which the most important are the *central moments* defined by:

$$M_n'[\xi] = \sum_{\forall i} (x_i - E[\xi])^n \cdot p_i$$

The second central moment, $M_2'[\xi]$, is the *variance* of ξ. The variance measures how far from the expected value, the values of the random variable are spread. In some sense it measures how likely the random variable is to be far from the mean. The actual value of the variance depends on the magnitude of the values of the random variables. In order to make the measure independent of scale, the *squared coefficient of variance* is often used. It is defined as follows:

$$C_\xi^2 = \frac{M_2'[\xi]}{E[\xi]^2} = \frac{M_2[\xi] - E[\xi]^2}{E[\xi]^2}$$

The constant random variable has coefficient of variance 0.

Factorial moments are defined by:

$$\tilde{M}_n[\xi] = \sum_{\forall i} \left(\prod_{j=0}^{n-1} (x_i - j) \right) p_i$$

It is clear that both central and factorial moments can be defined in terms of crude moments. For example,

$$\begin{aligned}
\tilde{M}_3[\xi] &= \sum_{\forall i} x_i(x_i - 1)(x_i - 2)p_i \\
&= \sum_{\forall i} (x_i^3 - 3x_i^2 + 2x_i)p_i \\
&= M_3[\xi] - 3M_2[\xi] + 2M_1[\xi]
\end{aligned}$$

In general, so long as one can calculate the expectation, the calculation of factorial or central moments will allow calculation of the crude moments.

2.3.1 Discrete distributions

Let us discuss some important discrete probability distributions. A *Bernoulli trial* is an experiment which has two possible outcomes. Arbitrarily we call them success, S, and failure F. Conventionally, success occurs with probability p and failure with probability $q = 1 - p$, on any trial independently of any other trials. Consider sequences of n Bernoulli trials. Clearly there are 2^n different sequences. We assign each event a probability equal to $p^k q^{n-k}$, where there are k successes. There are $\binom{n}{k}$ such events, so the probability of there being k successes in n trials independent of the order in which the successes occur, is:

$$\pi_k = \binom{n}{k} p^k q^{n-k}$$

We can see that the set $\{\pi_k\}$ is a probability distribution since:

$$\sum_{i=0}^{n} \pi_i = \sum_{i=0}^{n} \binom{n}{i} p^i q^{n-i}$$
$$= (p + q)^n$$

which is trivially equal to 1. This distribution is called the *binomial distribution*, for obvious reasons. The expectation of a binomial distribution is given by:

$$E[\xi] = \sum_{i=0}^{n} i \cdot \binom{n}{i} p^i q^{n-i} = np$$

and the variance similarly is given by $np(1 - p)$.

Other distributions can be observed when infinite sequences of Bernoulli trials are considered. The points in our sample space are infinite sequences of S and F, such as:

$$SSSSSSSSSFFSSSFFFSFSFFSSSF\cdots$$
$$FSSSSSFSFFFSSSFFSSSFFFSFSF\cdots$$
$$SSFFFSFFSFFFSSSSFSSFFFSFSS\cdots$$

We cannot assign probabilities to these sample points directly, since there are a countably infinite number of sample points. However, if we define random variables based on these sample points, it becomes possible to find the probabilities of the random variable taking certain values. If we define the random variable ζ as the number of trials until the first failure, then it will take on the values 9, 0, and 2, in the above samples. The value of ζ will be i for that subset of sample points in which i successes occur, followed by a single failure. We assign probabilities to those subsets of events. Since the trials are independent, each success occurs with probability p and the failure occurs with probability q. We define:

$$\Pr(\zeta = i) = p^i q$$

The only event which has not had a probability assigned to it under this mapping is the event consisting of an infinite number of successes, and no failures. However, this assignment of probability values is a distribution, since:

$$\sum_{i=0}^{\infty} \Pr(\zeta = i) = q \sum_{i=0}^{\infty} p^i$$
$$= q \cdot \frac{1}{1-p} = 1$$

Thus the probability of a continuous run of successes and no failures is 0. The expected number of successes before the first failure is given by:

$$\sum_{i=0}^{\infty} i \Pr(\zeta = i) = q \sum_{i=0}^{\infty} i p^i$$
$$= qp \sum_{i=0}^{\infty} i p^{i-1}$$
$$= qp \sum_{i=0}^{\infty} \frac{\mathrm{d}}{\mathrm{d}p} p^i$$
$$= qp \frac{\mathrm{d}}{\mathrm{d}p} \sum_{i=0}^{\infty} p^i$$
$$= qp \cdot \frac{\mathrm{d}}{\mathrm{d}p} \frac{1}{1-p}$$
$$= qp \frac{1}{(1-p)^2}$$
$$= \frac{p}{1-p}$$

Similar manipulation gives the variance as $p/(1-p)^2$. This is known as the *geometric distribution*.

The geometric distribution is sometimes defined as the number of Bernoulli trials to give the first failure, including the failure in the count. The calculation of these probabilities, and the mean and variance of this distribution is left as an exercise for the reader. In this book, this distribution will be known as the *modified geometric distribution*.

Another commonly encountered distribution is the *negative binomial*. The random variable here is the number of successes that precede the kth failure. Proceeding analogously to above, we can define $P(n)$ as:

$$P(n) = \binom{n+k-1}{n} p^n q^k$$

2.4   Continuous random variables

For some sample spaces, discrete values of random variables are not appropriate. For example, if we are interested in the number of jobs which use the graph plotter each hour then a discrete random variable is ideal. Alternatively, the duration of a job's use of the plotter can take almost any non-negative value, so a continuous random variable is indicated. As in the case of the infinite sequences of Bernoulli trials, we cannot assign a probability to each sample point in a meaningful way. We define the probabilities in terms of the *distribution function*, which is a function giving the probability that the random variable is less than its argument. Some other authors refer to the distribution function as the cumulative distribution function to emphasise the fact that it includes the probabilities of all lesser values. If the distribution function of a random variable, ζ, is $F(x)$, then:

$$\Pr(\zeta \leq x) = F(x)$$

In order to be a probability distribution function, a function must satisfy four conditions:

1. $F(x_1) \leq F(x_2)$ if $x_1 \leq x_2$.
2. $F(x) \to 0$ as $x \to -\infty$.
3. $F(x) \to 1$ as $x \to \infty$.
4. $F(x)$ is right continuous.

Note that the function does not need to be continuous, allowing points which have a strictly positive probability of occurrence.[†] The derivative of the distribution function is known as the *density function*, where it exists:

$$f(x) = \frac{\mathrm{d}F(x)}{\mathrm{d}x}$$

It can be convenient to regard the infinitesimal quantity $f(x)\mathrm{d}x$ as the probability that the value of ζ lies in the interval $(x, x + \mathrm{d}x)$. This simplifies many arguments, without compromising their validity.

The expectation of a random variable, ζ, with distribution function $F(x)$ is:

$$E[\zeta] = \int_{-\infty}^{\infty} x \, \mathrm{d}F$$

where the integral is a Riemann-Stieltjes integral. If $F(x)$ is continuous with derivative $f(x)$, this is equal to the more familiar Riemann integral:

[†]Discrete random variables can be expressed in terms of their distribution functions too, but this makes the description more complex, and requires more advanced mathematics, such as measure theory.

$$E[\zeta] = \int_{-\infty}^{\infty} x f(x) \, \mathrm{d}x$$

If there are points which have a strictly positive probability, that is discontinuities in F, then the Riemann-Stieltjes integral can be evaluated by calculating the ordinary Riemann integral in the continuous intervals of F, and adding $x_i(F(x_i) - F(x_i^-))$ for each point of discontinuity x_i. It can be seen that a discrete probability distribution fits this pattern, since the distribution function is a step function, and the density function is 0 everywhere that it exists.

Moments of continuous random variables are defined analogously to those for discrete random variables:

$$M_n[\zeta] = \int_{-\infty}^{\infty} x^n \, \mathrm{d}F$$

and:

$$M_n'[\zeta] = \int_{-\infty}^{\infty} (x - E[\zeta])^n \, \mathrm{d}F$$

The definition of the squared coefficient of variation is identical to that for a discrete random variable.

The expectation of the sum of a number of random variables is the sum of their individual expectations. This is true even if the variables are dependent in some way. Consider the random variables, ξ and ζ, which have distribution function $F(x, y) = \Pr(\xi \le x \wedge \zeta \le y)$. The joint density function is $f(x, y) = \frac{\mathrm{d}^2 F(x,y)}{\mathrm{d}x \mathrm{d}y}$. The marginal distribution of ξ has density function $f_\xi(x) = \int_{-\infty}^{\infty} f(x, y) \, \mathrm{d}y$. What is the expectation of $\xi + \zeta$?

$$
\begin{aligned}
E[\xi + \zeta] &= \int_{-\infty}^{\infty} \int_{-\infty}^{\infty} (x + y) f(x, y) \, \mathrm{d}x \, \mathrm{d}y \\
&= \int_{-\infty}^{\infty} \int_{-\infty}^{\infty} x f(x, y) \, \mathrm{d}x \, \mathrm{d}y + \int_{-\infty}^{\infty} \int_{-\infty}^{\infty} y f(x, y) \, \mathrm{d}x \, \mathrm{d}y \\
&= \int_{-\infty}^{\infty} x f_\xi(x) \, \mathrm{d}x + \int_{-\infty}^{\infty} y f_\zeta(y) \, \mathrm{d}y \\
&= E[\xi] + E[\zeta]
\end{aligned}
$$

The independence of random variables is defined in an analogous manner to the independence of events. Two random variables, ξ and ζ, with distribution functions, $X(x)$ and $Z(x)$, are independent if their joint distribution function satisfies:

$$F(x, y) = P(\xi \le x \wedge \zeta \le y) = X(x) \cdot Z(y)$$

The independence of larger numbers of random variables is similarly defined.

This enables the distribution function of the maximum of a set of independent random variables to be found. Let θ be the random variable defined as the maximum of the random variables ξ and ζ above. What is the distribution function of θ?

$$\begin{aligned} \Pr(\theta \leq x) &= \Pr(\max(\xi, \zeta) \leq x) \\ &= \Pr(\xi \leq x \wedge \zeta \leq x) \\ &= X(x) \cdot Z(x) \end{aligned}$$

In particular, this means that the distribution function of the maximum of n independent samples of the random variable ξ has distribution function $X(x)^n$.

2.4.1 Continuous distributions

The *uniform distribution* corresponds to the discrete case of equally likely outcomes. A variable is uniformly distributed between a and b if its distribution function is:

$$F(x) = \begin{cases} 0, & x \leq a \\ \frac{x-a}{b-a}, & a \leq x \leq b \\ 1, & x > b \end{cases}$$

The density function, $f(x)$, is identically 0, except in the range $a \leq x \leq b$, where it is the constant $1/(b-a)$. The expected value is $(a+b)/2$, and the variance is $(b-a)^2/12$.

The most important continuous distribution for performance modelling purposes is the *exponential distribution*, which has density function:

$$f(x) = \mu e^{-\mu x}, \quad x \geq 0$$

It has distribution function $F(x) = 1 - e^{-\mu x}$. It has mean $1/\mu$, and variance $1/\mu^2$. The squared coefficient of variation is 1. Some authors call this distribution the *negative exponential* distribution.

The normal distribution is of great use to statisticians because the distribution of the sum of a large number of independent random variables tends to a normal distribution as the number of variables increases. A *normal distribution*, sometimes known as a *Gaussian distribution*, with mean μ and variance σ^2, has density function:

$$f(x) = \frac{1}{\sqrt{2\pi}\sigma} e^{-(x-\mu)^2/\sigma^2}, \quad -\infty < x < \infty$$

2.5 Generating functions and Laplace transforms

Although we can often derive the full set of probability values for a discrete distribution, if the sample space is infinite, we have a rather unwieldy set of formulas to manipulate. A convenient tool for manipulating infinite series is the generating function.

If the values $p_0, p_1, \ldots$ form a probability distribution, then the distribution's *probability generating function* is given by:

$$G(z) = E[z^i] = \sum_{i=0}^{\infty} z^i p_i$$

The convergence of the series is guaranteed for all z less than or equal to 1, since the probabilities must sum to 1. That property also ensures that $G(1) = 1$. Unless otherwise stated, all generating functions used in this book are probability generating functions. The probability generating function is sometimes called the *z-transform*. The generating function contains all the information that was in the individual point probabilities. In principle, we can extract the value of any probability using repeated differentiation, since:

$$p_j = \frac{1}{j!} \left. \frac{d^j G(z)}{dz^j} \right|_{z=0}$$

More importantly, the moments of the distribution can also be calculated from the generating function. For example, the expectation is given by:

$$
\begin{aligned}
E[\xi] &= \sum_{i=0}^{\infty} i p_i \\
&= \sum_{i=0}^{\infty} i p_i z^{i-1} \quad \text{evaluated at } z = 1 \\
&= \left. \frac{d}{dz} \sum_{i=0}^{\infty} p_i z^i \right|_{z=1} \\
&= G'(1)
\end{aligned}
$$

Similarly, the second moment is given by:

$$
\begin{aligned}
E[\xi^2] &= \sum_{i=0}^{\infty} i^2 p_i \\
&= \sum_{i=0}^{\infty} i(i-1) p_i + \sum_{i=0}^{\infty} i p_i \\
&= \sum_{i=0}^{\infty} i(i-1) p_i z^{i-2} + \sum_{i=0}^{\infty} i p_i z^{i-1} \quad \text{evaluated at } z = 1 \\
&= G''(1) + G'(1)
\end{aligned}
$$

In general, the kth derivative of the generating function, evaluated at 1, is the kth factorial moment of the random variable. It is intuitively obvious that there is a one to one correspondence between distributions and their probability generating functions. If two generating functions are identical, then the underlying distributions must also be identical.

The generating function of the geometric distribution is given by:

$$G(z) = \sum_{i=0}^{\infty} (1-p)p^i z^i$$
$$= \frac{1-p}{1-pz}$$

Use of the generating function to calculate the moments is usually much simpler than the sort of trickery employed in section 2.3.

One of the most useful properties of the generating function is demonstrated in the calculation of the generating function of the sum of a number of independent random variables. Consider two discrete random variables, ξ and ζ, with probability distributions p_i and q_i, and generating functions $X(z)$ and $Z(z)$, respectively. Let the distribution of $\sigma = \xi + \zeta$ be s_i, with generating function $S(z)$. $\sigma = 0$ only if both ξ and ζ equal 0, which occurs with probability $p_0 \cdot q_0$. For σ to equal 1, either $\xi = 1$ and $\zeta = 0$, which occurs with probability $p_1 \cdot q_0$, or $\xi = 0$ and $\zeta = 1$, with probability $p_0 \cdot q_1$. Hence $s_1 = p_1 q_0 + p_0 q_1$. In general, $s_k = \sum_{i=0}^{k} p_i q_{k-i}$. This is just the rule for calculating the coefficients of the product of two power series. The generating function of σ, $S(z) = X(z) \cdot Z(z)$.

There are a number of transforms which can be used to manipulate the distribution functions of continuous random variables. The moment generating function, characteristic function, and Laplace-Stieltjes transform all have similar convenient properties. We shall only consider the *Laplace-Stieltjes transform,* or Laplace transform, and restrict consideration to non-negative random variables. A non-negative random variable, τ, with distribution function $F(t)$, has a Laplace transform denoted by $F^*(s)$,

$$F^*(s) = E[e^{-s\tau}]$$
$$= \int_0^{\infty} e^{-st} \, dF(t)$$
$$= \int_0^{\infty} e^{-st} f(t) \, dt$$

Use of the Riemann-Stieltjes integral means that distributions with discontinuous distributions have well defined transforms. For example, if there is a discontinuity at t_0, then the Laplace transform of the distribution $F(t)$ is calculated by:

$$F^*(s) = \int_0^{t_0^-} e^{-st} f(t) \, dt + e^{-st_0}(F(t_0) - F(t_0^-)) + \int_{t_0}^{\infty} e^{-st} f(t) \, dt$$

The moments of the distribution can be retrieved from the Laplace transform. The expectation is given by:

$$
\begin{aligned}
E[\tau] &= \int_0^\infty t f(t)\,\mathrm{d}t \\
&= \int_0^\infty t e^{-st} f(t)\,\mathrm{d}t \Big|_{s=0} \\
&= -\frac{\mathrm{d}}{\mathrm{d}s} \int_0^\infty e^{-st} f(t)\,\mathrm{d}t \Big|_{s=0} \\
&= -F^{*\prime}(0)
\end{aligned}
$$

Similarly, one can derive the second moment as $E[\tau^2] = F^{*\prime\prime}(0)$, and so on. The Laplace transform has a similar property to generating functions. There is a one to one correspondence between distributions and their Laplace transforms. The transform of the sum of two independent random variables is the product of their transforms. Consider two random variables, ϕ and γ. The Laplace transform of their sum, η, is given by:

$$
E[e^{-s\eta}] = E[e^{-s(\phi+\gamma)}] = E[e^{-s\phi}e^{-s\gamma}] = E[e^{-s\phi}] \cdot E[e^{-s\gamma}]
$$

Extraction of the distribution function, rather than its moments, from the transform is more difficult than extracting individual state probabilities from a generating function. Inversion of arbitrary Laplace transforms is well defined and involves integrating the transform round a contour in the complex plane. The simplest technique, however, is to try to identify similar Laplace transforms of a similar form in a table of transforms, and hence deduce the form of the distribution function.

The exponential distribution with parameter μ has density function $f(t) = \mu e^{-\mu t}$. Its Laplace transform is given by:

$$
\int_0^\infty f(t) e^{-st}\,\mathrm{d}t = \mu \int_0^\infty e^{-(s+\mu)t}\,\mathrm{d}t = \frac{\mu}{s+\mu}
$$

2.6 Exercises

1. Consider a pack of playing cards. Assume that it has been well shuffled. What is the probability that the top two cards are both hearts? What is the probability that they are of the same suit? What is the probability that they are a pair? What is the probability that they are a pair of aces?

2. What is the probability of *one* player at bridge being dealt a hand containing all 13 cards from one suit?

3. If a player examines his hand after six cards have been dealt, and discovers that all his cards are hearts, what is the probability that he will be dealt the remaining seven cards in the suit?

4. What is the probability that a player will have no hearts in his hand after 8 cards have been dealt?

5. Assuming that the sex of a child is a Bernoulli trial with $p = 1/2$, calculate the probability that a family with three children had a boy first, and then two girls. What is the probability that in a family of three, the eldest child is a boy?

6. My brother has two children. What is the probability that he has a son? What is the probability that he has only one son? What is the probability that he has no sons?

7. If all 36 numbers on a roulette wheel are equi-probable, what is the probability that 100 spins of the wheel will not produce a single '17'?

8. Evaluate the mean and variance of the modified geometric distribution. Calculate its probability generating function.

9. Calculate the Laplace transform of the uniform distribution.

10. What is the distribution function of the minimum of n independent samples of the random variable ξ, if the distribution function of ξ is $F(x)$?

2.7 Further reading

Probability theory has a vast literature and there are many introductory texts which cover the material of this chapter in greater depth. The classic text by Feller[7, 6] is still available and a convenient reference. Less heavyweight introductions are to be found in such texts as Beaumont[1, 2]. Ross[10] is another text which is popular. Chung[4] is a more mathematical treatment. Blake[3], Drake[5], and Haight[8] are all concerned with applications of probability theory. Johnson and Kotz have published a guide to various distributions[9].

2.8 Bibliography

[1] G.P. Beaumont. *Introductory Applied Probability*. Ellis Horwood, Chichester, 1983.

[2] G.P. Beaumont. *Probability and Random Variables*. Ellis Horwood, Chichester, 1986.

[3] I.F. Blake. *An Introduction to Applied Probability*. Wiley, New York, NY, 1979.

[4] K.L. Chung. *A Course in Probability Theory*. Academic Press, London, second edition, 1974.

[5] A.W. Drake. *Fundamentals of Applied Probability Theory*. McGraw-Hill, New York, NY, 1967.

[6] W. Feller. *An Introduction to Probability Theory and Its Applications*, volume 1. Wiley, New York, NY, third edition, 1968.

[7] W. Feller. *An Introduction to Probability Theory and Its Applications*, volume 2. Wiley, New York, NY, second edition, 1971.

[8]   F.A. Haight. *Applied Probability*. Plenum Press, New York, NY, 1981.

[9]   N.L. Johnson and S. Kotz. *Discrete Distributions*. Wiley, New York, NY, 1969.

[10] S.M. Ross. *A First Course in Probability*. Macmillan, London, 1976.

Chapter 3

Stochastic processes

Informally, a stochastic process is a random process which is changing in time. For example, the number of passengers waiting in a bus queue is a stochastic process. It increases when new passengers join the queue, and may decrease when a bus arrives. Formally, a stochastic process is defined as a set of random variables $\{\zeta(t)|t \in T\}$, for some index set T. The set T is usually interpreted as a set of time instants. When T is countable, we have a discrete-time stochastic process; when the elements of T are a continuous set of real numbers, the process is a continuous-time stochastic process. $\zeta(t)$ is called the state of the process at time t. A *sample path* is a single instance of the stochastic process. Arguments which apply to any sample path, such as our proof of Little's theorem in chapter 1, apply to the process as a whole too. A stochastic process is said to be *stationary* if the distribution of $\zeta(t + s) - \zeta(t)$ is independent of t.

3.1 Poisson process

A Poisson process is one of the simplest interesting stochastic processes. It can be defined in a number of ways, which are equivalent. The following axiomatic definition seems most appropriate to our needs. A *Poisson process* is a counting process that records the number of events which occur in an interval. A Poisson process with rate λ satisfies the following properties:

1. $\Pr$(a single event in an interval of duration δt) $= \lambda \delta t + o(\delta t)$, independently of when the last event occurred, and of all other intervals.
2. $\Pr$(More than one event will occur in an interval of duration δt) $= o(\delta t)$.

From the axioms of probability, the probability that no events occur in the interval of length δt is $1 - \lambda \delta t + o(\delta t)$.

From this definition, we can derive $P_n(t)$, the probability of n events in an interval of length t. Consider the interval divided into m equal subintervals ($m > n$). Either the events occur so that n of the intervals have a single event, and $m - n$ have no events, or there are some intervals with more than one event. Denoting the first possibility by $P_n^{(a)}(t)$ and the second by $P_n^{(b)}(t)$, since these two possibilities are mutually exclusive, we have:

$$P_n(t) = P_n^{(a)}(t) + P_n^{(b)}(t)$$

Consider $P_n^{(a)}(t)$, and choose the n subintervals in advance; then the probability of this occurrence is:

$$\text{Pr(given } n \text{ intervals have 1 event each and remaining } m - n \text{ have none)}$$
$$= \left(\lambda\frac{t}{m}\right)^n \left(1 - \lambda\frac{t}{m} + o(\frac{t}{m})\right)^{m-n}$$

The choice of the n intervals can be made in $\binom{m}{n}$ distinct ways. Hence:

$$P_n^{(a)}(t) = \frac{m!}{(m-n)!n!}\left(\frac{\lambda t}{m}\right)^n \left(1 - \frac{\lambda t}{m} + o(\frac{t}{m})\right)^{m-n}$$
$$= \frac{(\lambda t)^n}{n!} \cdot \frac{m!}{(m-n)!m^n} \cdot \left(1 - \frac{\lambda t}{m} + o(\frac{t}{m})\right)^{-n}$$
$$\cdot \left(1 - \frac{\lambda t}{m}\right)^m \cdot \left(1 + o(\frac{t}{m})\right)^m$$

Now we take the limit as $m \to \infty$ and find:

$$P_n^{(a)}(t) = \frac{(\lambda t)^n}{n!} \cdot 1 \cdot 1 \cdot \mathrm{e}^{-\lambda t} \cdot 1 = \frac{(\lambda t)^n}{n!}\mathrm{e}^{-\lambda t}$$

Recalling that $P_n(t) = P_n^{(a)}(t) + P_n^{(b)}(t)$ and that since $P_n(t)$ is a probability distribution, $\sum_{n=0}^{\infty} P_n(t) = 1$,

$$\sum_{n=0}^{\infty} P_n(t) = \sum_{n=0}^{\infty} P_n^{(a)}(t) + \sum_{n=0}^{\infty} P_n^{(b)}(t)$$
$$= \mathrm{e}^{-\lambda t} \sum_{n=0}^{\infty} \frac{(\lambda t)^n}{n!} + \sum_{n=0}^{\infty} P_n^{(b)}(t)$$
$$= 1 + \sum_{n=0}^{\infty} P_n^{(b)}(t)$$

Hence $P_n^{(b)}(t) = 0$, since no probability can be negative, and:

$$P_n(t) = \frac{(\lambda t)^n}{n!}\mathrm{e}^{-\lambda t}$$

which is the Poisson distribution with parameter λt.

The mean number of events in an interval of length t is given by:

$$E[n(t)] = \sum_{k=1}^{\infty} kP_k(t) = \lambda t$$

The generating function of the number of events in an interval of length t is $G(z) = e^{\lambda t(z-1)}$.

The time interval between successive events can also be deduced. Let $F(t)$ be the distribution function of τ, the time between events. Consider $\Pr(\tau > t) = 1 - F(t)$:

$$\Pr(\tau > t) = \Pr(\text{No events in an interval of length } t)$$
$$= e^{-\lambda t}$$

Hence:

$$F(t) = 1 - e^{-\lambda t}$$

So the interevent time distribution is an exponential distribution. The Poisson and exponential distributions have a special place in queueing theory and performance modelling because of their 'memoryless' property. It is clear from the definition of the Poisson process that the number of events in distinct intervals is dependent only on the interval lengths, and not on the behaviour of the event process in preceding or subsequent intervals. The *memoryless property* of the exponential distribution is so called because the time to the next event is independent of when the last event occurred. Suppose that the last event was at time 0, what is the probability that the next event will be after $t + s$, given that time t has elapsed since the last event, and no events have occurred?

$$\Pr(\tau > t + s | \tau > t) = \frac{\Pr(\tau > t + s \text{ and } \tau > t)}{\Pr(\tau > t)}$$
$$= \frac{e^{-\lambda(t+s)}}{e^{-\lambda t}}$$
$$= e^{-\lambda s}$$

Thus the knowledge that there have been no events from a Poisson stream for a period of time does not alter the distribution of time until the next event. Conversely, the knowledge that there has been a large number of events in the recent past does not change the distribution of time until the next event.

3.2   Random walk

A *random walk*, or drunkard's walk, is a stochastic process in which the states are integers, X_n representing the position of a particle at time n. At each (discrete) time interval, X_n can either increase or decrease by 1. Each state change is independent of the current position, and of previous state changes:

$$X_n = X_{n-1} + Z_n$$

where Z_n is a random variable which takes values 1 or -1, independently of n. If $X_0 = 0$, then $X_n = \sum_{i=1}^{n} Z_n$. For fixed n, this is clearly related to the binomial distribution. The value of X_n is the difference between the number of successes and failures in n Bernoulli trials, where success in the ith trial is represented by $Z_i = 1$ and failures by $Z_i = -1$.

Among the properties that have been studied for the random walk are the *first passage time* which is the time until a particular state is reached for the first time and the *first return time* which is the time until the starting state is reentered. The simple random walk can be extended in several ways: the values that Z_i can take may include 0; absorbing barriers may be erected, so that the process terminates on reaching some value; reflecting barriers restrict the values of X_n to prevent the process escaping from some set of positions.

3.3 Markov processes and chains

A *Markov process* is a stochastic process in which the distribution at any time in the future depends only on the current state of the process, and *not* on how that state was reached. Formally, a stochastic process, $\zeta(t)$, is a Markov process if:

$$\begin{aligned}
&\Pr(\zeta(t) = x | \zeta(t_n) = x_n, \zeta(t_{n-1}) = x_{n-1}, \ldots, \zeta(t_0) = x_0) = \\
&\quad \Pr(\zeta(t) = x | \zeta(t_n) = x_n), \quad \text{for } t_0 < t_1 < \ldots < t_n < t
\end{aligned} \tag{3.1}$$

The set of values assumed by $\zeta(t)$, the *state space*, can be discrete or continuous. The elements of the index set, t, can be discrete or continuous also. It is possible for these conditional probabilities to depend on time, but *time-homogeneous* processes, in which the conditional probabilities are independent of time, are much more common in practical modelling situations.

3.3.1 Markov chains

A Markov process with discrete state space and discrete index set is called a *Markov chain*, and the necessary condition that the process, X_n, must satisfy can be written as:

$$\begin{aligned}
&\Pr(X_{n+1} = m | X_n = x_n, X_{n-1} = x_{n-1}, \ldots, X_0 = x_0) = \\
&\quad \Pr(X_{n+1} = m | X_n = x_n)
\end{aligned}$$

Again, these probabilities can depend on n. The future behaviour of a Markov chain depends only on its current state, and not on how that state was reached. This is the Markov, or memoryless, property.

Given that a chain is in state i at time n, then the one-step *transition probabilities* are the values $p_{ij}(n)$:

$$p_{ij}(n) = \Pr(X_{n+1} = j | X_n = i)$$

Usually, we shall deal with chains with stationary transition probabilities. For these *homogeneous* Markov chains, the dependence on n will be omitted and we write the one-step transition probabilities as p_{ij}. Notice that the one-step transition probabilities give all the information we need to predict the (probabilistic) behaviour of the process at all future times. For example, if the process is in state i at time n, what is the probability that it will be in state k at time $n+2$. Denote this two-step probability by $p_{ik}^{(2)}$, and observe that to reach k in two steps, the process must make a transition to some state j in one step with probability p_{ij}, (j may equal i or k), and then make the transition to state k, with probability p_{jk}. The two-step transition probability is given by summing over all possible intermediate states:

$$p_{ik}^{(2)} = \sum_j p_{ij} \cdot p_{jk}$$

The one-step transition probabilities can be written in the form of a matrix, $P = (p_{ij})$. The *transition matrix* is a *stochastic matrix*, that is, the sum of the elements in each row is 1. Writing $\mathbf{1}$ for the column vector equal to 1 in every component:

$$P\mathbf{1} = \mathbf{1}$$

Premultiplying each side of the relationship by P:

$$P \cdot P\mathbf{1} = P^2\mathbf{1} = P\mathbf{1} = \mathbf{1}$$

it is seen that any power of a stochastic matrix is itself stochastic. This property holds, even if the state space is countably infinite. Writing the state probabilities at time n as a row vector, $\boldsymbol{\pi}(n) = (\pi_0(n), \pi_1(n), \ldots)$. The evolution of the process can then be expressed as:

$$\boldsymbol{\pi}(n+1) = \boldsymbol{\pi}(n)P$$

This relationship can be iterated, and we find that:

$$\boldsymbol{\pi}(n) = \boldsymbol{\pi}(0)P^n$$

so that the initial probability distribution, $\boldsymbol{\pi}(0)$, and the transition matrix, P, are sufficient to calculate the state probabilities at all times in the future. The two-step transition probabilities can be expressed in matrix form as P^2. In general, the n-step transition matrix is given by P^n. The probabilistic expression of this fact is the *Chapman-Kolmogorov equation* for homogeneous Markov chains:

$$p_{ij}^{(m+n)} = \sum_k p_{ik}^{(m)} p_{kj}^{(n)}$$

3.3.2   State classification

A state j is said to be *accessible* from a state i, if $p_{ij}^{(n)} > 0$, for some $n > 0$. If state i is accessible from state j, and state j is accessible from state i, then states i and j are said to *communicate*. It is clear that if states i and j communicate, and states j and k communicate, then states i and k also communicate. States which communicate are in the same communicating class. If all states of the Markov chain belong to the same communicating class, then the chain is said to be *irreducible*.

A state, i, is said to have period, d, if $p_{ii}^{(n)} = 0$ for all n that are not divisible by d, and d is the largest integer with this property. A state with period 1 is said to be *aperiodic*. All states in the same communicating class have the same period. If the chain is irreducible, the it can be referred to as periodic or aperiodic according to the classification of its states.

A state is said to be *recurrent*, if it will be entered again with probability 1. If the probability of return to the state is less than 1, then the state is *transient*. Recurrent states are further classified as *positive recurrent* if the mean number of steps until the state is returned to is finite, and as *null-recurrent* if the mean number of steps until the state is returned to is infinite. An irreducible, aperiodic Markov chain of which all states are positive-recurrent is an *ergodic* Markov chain. The expected number of visits to a transient state is finite. It follows that a chain with only a finite number of states must have at least one recurrent state.

3.3.3   Stationary distribution

A probability distribution $\boldsymbol{\pi} = (\pi_1, \pi_2, \ldots)$ is a *stationary distribution* of the Markov chain if it satisfies:

$$\boldsymbol{\pi} = \boldsymbol{\pi} P \tag{3.2}$$

For ergodic Markov chains, it can be proved that there exists a unique stationary distribution and that the limiting distribution, as the number of steps becomes large, is the stationary distribution, independent of the starting state.

3.3.4   Markov processes

The Markov process with a discrete state space, but continuous time is the other type of stochastic process with which we shall be concerned. It is possible to have Markov processes with continuous state spaces, and they have some applications in modelling, but the added complications make them a specialised topic.

A (stationary, n state) Markov process can be characterised by its ($n \times n$) *infinitesimal generator matrix* $Q = (q_{ij})$. If the state of the system at time t is $X(t)$, then:

$$\Pr(X(t + \delta t) = j | X(t) = i) = q_{ij}\delta t + o(\delta t), \quad i \neq j$$

State classification is identical to that for Markov chains. We shall assume that the process is ergodic, that is every state can eventually be reached from every other state. The presence of transient states does not usually cause any problems, but if there is more than one class of communicating states difficulties can arise. This is not usually a problem in practical modelling situations. If we denote the probability that the system is in state k at time t by $\pi_k(t)$, then the distribution a short time δt later is given by:

$$\pi_k(t + \delta t) = \pi_k(t)(1 - \sum_{j \neq k} q_{kj}\delta t) + (\sum_{i \neq k} q_{ik}\pi_i(t))\delta t + o(\delta t)$$

Defining $q_{ii} = -\sum_{j \neq i} q_{ij}$:

$$\pi_k(t + \delta t) = \pi_k(t) + (\sum_{i=1}^{n} q_{ik}\pi_i(t))\delta t + o(\delta t) \qquad (3.3)$$

which upon rearrangement and taking the limit as $\delta t \to 0$ gives:

$$\frac{d\pi_k}{dt} = \sum_{i=1}^{n} q_{ik}\pi_i(t)$$

Denoting the row vector $(\pi_1(t), \pi_2(t), \ldots, \pi_n(t))$ by $\boldsymbol{\pi}(t)$ these equations can be expressed in matrix notation as:

$$\frac{d\boldsymbol{\pi}(t)}{dt} = \boldsymbol{\pi}(t)Q$$

When steady state is reached, $\frac{d\boldsymbol{\pi}}{dt} = 0$, so:

$$\boldsymbol{\pi}Q = \mathbf{0} \qquad (3.4)$$

where $\boldsymbol{\pi}$ is the steady state distribution vector. Although this presentation has been given in terms of a finite state space, infinite state space processes can also be expressed in terms of infinite dimensional infinitesimal generator matrices. These are well defined so long as the infinite sums that arise in calculating vector products are well defined.

3.3.5   Local balance and time reversibility

Some Markov processes satisfy a particularly convenient property known as *local balance*. In a process with local balance, the probability flow from state i to state j equals the probability flow from state j to state i, that is:

$$\pi_i q_{ij} = \pi_j q_{ji} \qquad (3.5)$$

If we consider equation (3.3) in steady state, then we have:

$$\sum_{j=1}^{n} q_{ji}\pi_j = 0$$

Summing equation (3.5) over all $j \neq i$, we have:

$$\pi_i \sum_{j \neq i} q_{ij} = \sum_{j \neq i} q_{ji}\pi_j$$

where the sum on the left-hand side equals $-q_{ii}$, by definition. Hence the global balance equations are the sum of the local balance equations. If a solution to the local balance equations can be found, then it will also be a solution to the global balance equations. Notice that many Markov processes do not satisfy local balance. In particular, before there can be local balance, for every state i, any state that has a non-zero transition rate from state i must also have a non-zero transition rate to state i.

Local balance is intimately connected with the *time-reversibility* property of some Markov processes. Probabilistically, a reversible Markov process is statistically identical to itself when viewed in reverse time. In a queueing situation, that implies that every arrival in the original process is viewed as a departure from the reversed process, and every departure from the original process is an arrival in the reversed process. Many interesting results are available to assist in the analysis of time reversible processes. Their local balance property makes many solutions have a particularly simple form.

3.4   Renewal theory

One of the most general stochastic processes for which interesting and useful results can be developed is the *renewal process*. A renewal process is a process that counts the occurrences of events when the time between successive events is an independent sample from some arbitrary distribution. If the distribution of the time between event $i - 1$ and event i has probability density function $F_i(t)$, and T_1, T_2, ..., are independent samples from the appropriate distributions, then the event times are $S_i = \sum_{j=1}^{i} T_j$. The distribution of S_r is denoted by $R_r(t)$. If the mean of the distribution of T_j is τ, then $E[S_r] = r\tau$. The distribution of the number of events, n_t, that occur before t is related to this, because:

$$\Pr(n_t < r) = \Pr(S_r > t) = 1 - R_r(t)$$

It follows that:

$$\Pr(n_t = r) = R_r(t) - R_{r+1}(t)$$

The expected number of renewals before t is given by:

$$N(t) = E[n_t] = \sum_{r=1}^{\infty} r(R_r(t) - R_{r+1}(t)) = \sum_{r=1}^{\infty} R_r(t)$$

The Laplace transform of this can be taken:

$$N^*(s) = E[e^{-sE[n_t]}] = \sum_{r=1}^{\infty} R_r^*(s)$$

where $R_r^*(s)$ is the Laplace transform of $R_r(t)$. By the convolution properties of the Laplace transform, $R_r^*(s) = \prod_{i=1}^{r} F_r^*(s)$. If the $F_i(t)$ are identical for $i \geq 2$, with Laplace transform $F^*(s)$, then $R_r^*(s) = F_1^*(s) \cdot (F^*(s))^{r-1}$, and:

$$N^*(s) = \frac{F_1^*(s)}{s(1 - F^*(s))}$$

Choosing $F_1(t)$ so that:

$$\frac{\mathrm{d}F_1(t)}{\mathrm{d}t} = \frac{1 - F(t)}{\tau}$$

the Laplace transform of $N(t)$ becomes:

$$N^*(s) = \frac{1}{\tau s^2}$$

which is easily inverted to give:

$$N(t) = \frac{t}{\tau}$$

The elementary renewal theorem states that, independent of the choice of $F_1(t)$ above,

$$\frac{N(t)}{t} \rightarrow \frac{1}{\tau} \quad \text{as } t \rightarrow \infty$$

A *regenerative process* is a stochastic process which repeats itself probabilistically. That is, there exists a sequence of regeneration points, $T_1, T_2, \ldots$, such that the process starting at T_2 ($T_3, \ldots$) is a probabilistic replica of the process starting at T_1. The simple random walk studied in section 3.2 is a regenerative process, with regeneration points every time the walk returns to 0. The time between regeneration points is known as a *cycle*. The fundamental theorem of regenerative processes, which we state without proof, allows us to derive the long run probabilities that the system is in some state, j say, using the expectations of the time it spends in state j during a cycle.

$$\pi_j = \frac{E[\text{time spent in } j \text{ in a cycle}]}{E[\text{length of a cycle}]}$$

3.5 Proof of Little's theorem

The proof of Little's theorem given in chapter 1 depends on the system emptying infinitely often, so that the contribution of the currently resident customers to W_t becomes negligible when the limiting process is performed. This is not necessary, and we now give a proof which does not require this renewal property. It holds for *any* sample path, that is, for any particular instantiation of the random variables, as well as for the 'average' case.

Let customers be numbered in order of arrival in the system, and denote the time of arrival of customer n by t_n. Then $0 \leq t_1 \leq t_2 \leq \ldots$. Let the residence time of the nth customer be W_n. Define the function $I_n(t)$ as 1 if customer n is in the system, and as 0 otherwise. Clearly, $I_n(t) = 0$ for $t < t_n$ and for $t > t_n + W_n$. For $t_n \leq t \leq t_n + W_n$, $I_n(t) = 1$. Also define $N(t)$ as the number of customers in the system at time t. It follows that:

$$W_n = \int_0^\infty I_n(t)\,dt \tag{3.6}$$

$$N(t) = \sum_{n=1}^\infty I_n(t) \tag{3.7}$$

We define $U(t)$ as the sum of the waiting times of those customers that arrived in the interval $[0, t]$, and $V(t)$ as the sum of the waiting times of those customers that departed in the same interval.

$$U(t) = \sum_{n:t_n \leq t} W_n, \quad V(t) = \sum_{n:t_n+W_n \leq t} W_n, \quad t \geq 0$$

We have the following lemma.

Lemma 3.5.1 For all $T \geq 0$,

$$U(T) \geq \int_0^T N(t)\,dt \geq V(T)$$

Proof. Using (3.7),

$$\int_0^T N(t)\,dt = \int_0^T \sum_{n=1}^\infty I_n(t)\,dt$$

$$= \sum_{n=1}^\infty \int_0^T I_n(t)\,dt$$

$$= \sum_{n:t_n \leq T} \int_0^T I_n(t)\,dt$$

From the relationship between W_n and $I_n(t)$ given by equation (3.6) it follows that:

$$\int_0^T I_n(t)\, dt \le W_n, \quad \text{for all } n \ge 1$$

with equality for those n such that $t_n + W_n \le T$. Hence,

$$\int_0^T N(t)\, dt \le U(T)$$

Also, since $\{n : t_n + W_n \le T\} \subseteq \{n : t_n \le T\}$,

$$\int_0^T N(t)\, dt \ge V(T)$$

The lemma can be explained intuitively as a method of accounting. Each customer costs the system an amount equal to his waiting time W_n. The three quantities in the inequality correspond to the cumulative cost to the system at time T if it is charged in one of three ways. Either in a single amount when the customer arrives ($U(T)$), or steadily at a rate of 1 unit per unit of waiting time ($\int_0^T N(t)\, dt$), or in a single amount when the customer departs ($V(T)$).

We now define $t_0 = 0$, and $A(t) = \max\{n | t_n \le t\}$, the cumulative number of arrivals in $[0, t]$. The following quantities are defined where the appropriate limits exist:

$$N = \lim_{T \to \infty} \frac{1}{T} \int_0^T N(t)\, dt \quad \lambda = \lim_{T \to \infty} \frac{A(T)}{T} \quad W = \lim_{M \to \infty} \frac{1}{M} \sum_{n=1}^M W_n$$

These are the definitions of the mean queue length, mean arrival rate, and mean waiting time, respectively. We can now prove Little's theorem.

Theorem 3.5.2 (Little)   If $\lambda < \infty$ and $W < \infty$, then $N < \infty$ and $N = \lambda W$.

Proof. t_n is well defined and finite for all n. Since λ is finite, $A(T) \to \infty$ as $T \to \infty$. It follows that $W = \lim_{T \to \infty} \sum_{n=1}^{A(T)} W_n / A(T)$. Since both λ and W are finite,

$$\lambda W = \lim_{T \to \infty} [A(T)/T] \cdot [1/A(T)] \sum_{n=1}^{A(T)} W_n$$

$$= \lim_{T \to \infty} (1/T) \sum_{n=1}^{A(T)} W_n$$

$$= \lim_{T \to \infty} U(T)/T$$

Using lemma 3.5.1, if we can show that $U(T)/T$ and $V(T)/T$ have a common limit as $T \to \infty$, then the theorem will follow.

Since λ and W are finite, $W_n/t_n \to 0$ as $n \to \infty$. Choose an arbitrary $\epsilon > 0$. Then there is an M'such that $n > M$ implies $W_n < t_n\epsilon$. Choose T such that $A(T) > M$. Then:

$$U(T) \geq V(T) = \sum_{n:t_n+W_n\leq T} W_n$$

$$= \sum_{n\leq M:t_n+W_n\leq T} W_n + \sum_{n>M:t_n+W_n\leq T} W_n$$

$$\geq \sum_{n\leq M:t_n+W_n\leq T} W_n + \sum_{n>M:t_n(1+\epsilon)\leq T} W_n$$

$$= \sum_{n\leq M:t_n+W_n\leq T} W_n + \sum_{n:t_n\leq T/(1+\epsilon)} W_n - \sum_{n\leq M:t_n(1+\epsilon)\leq T} W_n$$

The first and third terms of the right-hand side are bounded, since there are only a finite number of customers involved in each sum. Furthermore, they are of order $o(T)$ as $T \to \infty$. Thus:

$$U(T) \geq V(T) \geq U(T/(1 + \epsilon)) + o(T)$$

and dividing by T and letting $T \to \infty$,

$$\lim_{T\to\infty} \frac{U(T)}{T} \geq \lim_{T\to\infty} \frac{V(T)}{T} \geq [1/(1 + \epsilon)] \lim_{T\to\infty} \frac{U(T)}{T}$$

since:

$$\lim_{T\to\infty} \frac{U(T/(1 + \epsilon))}{T} = \frac{1}{1 + \epsilon} \lim_{T\to\infty} \frac{U(T/(1 + \epsilon))}{T/(1 + \epsilon)} = \frac{1}{1 + \epsilon} \lim_{T\to\infty} \frac{U(T)}{T}$$

We can choose ϵ as small as we like, so the equality follows.

We state without proof a generalisation of Little's theorem to higher moments of queue lengths and waiting times. In the general case, a rather complex formula is available, but when specialised to the $M/G/c$ queue, the following relationship holds between the kth factorial moment of queue length, and the kth crude moment of waiting time:

$$\tilde{M}_k(N) = \lambda^k E[W^k]$$

Since factorial moments can be expressed in terms of crude moments, this relationship can be very useful in finding the moments of waiting time, given the moments of queue length, which can usually be obtained more easily.

3.6   Further reading

There are many books on the theory of stochastic processes of differing levels of mathematical rigour. Cox and Miller[10] is an old, but still useful, example. More modern, but perhaps more theoretical, are Çinlar[7] and Ross[14]. Bhat[2] and Wong and Hajek[17] approach the subject from an applied perspective. Chung[4], Kemeny and Snell[12, 13], and Barucha-Reid[1] confine their attention to Markov chains and processes. Clarke and Disney[8] and Chung[5] combine elementary probability theory with stochastic processes. Taylor and Karlin[16] is a general introduction to stochastic modelling.

Time reversibility has been extensively studied by Kelly[11]. Renewal theory is treated in an excellent monograph by Cox[9]. Çinlar has also published a tutorial article[6]. The proof of Little's theorem given in this chapter is that developed by Stidham[15]. The generalisation of Little's theorem to higher moments of queue lengths and waiting times is due to Brumelle[3].

3.7   Bibliography

[1]   A.T. Barucha-Reid. *Elements of the Theory of Markov Processes and Their Applications*. McGraw-Hill, New York, NY, 1960.

[2]   U.N. Bhat. *Elements of Applied Stochastic Processes*. Wiley, New York, NY, 1984.

[3]   S.L. Brumelle. A generalization of $L = \lambda W$ to moments of a queue length and waiting times. *Operations Research*, 20(6):1127–1136, November/December 1972.

[4]   K.L. Chung. *Markov Chains with Stationary Transition Probabilities*. Springer-Verlag, Berlin, 1967.

[5]   K.L. Chung. *Elementary Probability Theory with Stochastic Processes*. Springer-Verlag, Berlin, third edition, 1979.

[6]   E. Çinlar. *Introduction to Stochastic Processes*. Prentice-Hall, Englewood Cliffs, NJ, 1975.

[7]   E. Çinlar. Markov renewal theory: A survey. *Management Science*, 21(7):727–752, 1975.

[8]   A.B. Clarke and R.L. Disney. *Probability and Random Processes: A First Course with Applications*. Wiley, New York, NY, 1985.

[9]   D.R. Cox. *Renewal Theory*. Methuen, London, 1962.

[10] D.R. Cox and H.D. Miller. *The Theory of Stochastic Processes*. Chapman and Hall, London, 1965.

[11] F.P. Kelly. *Reversibility and Stochastic Networks*. Wiley, London, 1979.

[12]   J.G. Kemeny and J.L. Snell. *Finite Markov Chains*. Van Nostrand, Princeton, NJ, 1960.

[13]   J.G. Kemeny, J.L. Snell, and A.W. Knapp. *Denumerable Markov Chains*. Van Nostrand, Princeton, NJ, 1966.

[14]   S.M. Ross. *Stochastic Processes*. Wiley, New York, NY, 1983.

[15]   S. Stidham. A last word on $L = \lambda W$. *Operations Research*, 22(2):417–421, March/April 1974.

[16]   H.M. Taylor and S. Karlin. *An Introduction to Stochastic Modeling*. Academic Press, London, 1984.

[17]   E. Wong and B. Hajek. *Stochastic Processes in Engineering Systems*. Springer-Verlag, Berlin, 1985.

Chapter 4

Simple queues

Markov processes hold a special place in stochastic process theory. The simplicity of being able to ignore how a process reached its current state when calculating the probability that it will be in some particular state at some later time makes analysis of systems which can be described as Markov processes particularly appealing. Although this analysis is presented in terms of messages arriving at a communications channel and being transmitted, the same analysis can be applied to jobs arriving at a CPU and being processed, or even to customers arriving in a butcher's shop and buying meat.

4.1 $M/M/1$ queue

We start our study of queueing systems performance with the $M/M/1$ queue. Messages arrive at the channel in a Poisson stream at rate λ, i.e., the time between successive message arrivals is exponentially distributed. The exponential distribution has the memoryless or Markov property. The distribution of the time until the next arrival is independent of the time that has elapsed since the last arrival. It can be shown that the exponential distribution is the only distribution with this property. Because knowledge of when the last arrival occurred does not change the distribution of the time until the next arrival, no information about arrivals need be kept to describe the state of the system.

Each message transmission takes a time which is exponentially distributed in length, with mean $1/\mu$. Transmission time is independent of the number of messages in the system, and of the transmission times of other messages. The distribution of the time until the next departure depends only on whether or not there is a message being transmitted and not on the length of time that the current transmission, if any, has lasted. The number of messages present in the system, both waiting and being transmitted, gives all the information needed to calculate the probabilities of numbers of messages at any time in the future (or the past!). Intuitively, the rate at which messages can be transmitted must be less than or equal to the rate at which they arrive if the queue is not to grow without bound. If the average arrival rate equals the average transmission rate then an exact analysis of the queue is a problem which requires advanced concepts of stochastic process theory. Since that case is rather pathological, only systems with the arrival rate strictly less than transmission rate will be analysed.

Whatever the state of the system, a new message will arrive in the next short interval δt with probability $\lambda \delta t$. If the system is not empty, the message being transmitted will finish transmission in the next interval of length δt with probability $\mu \delta t$; knowledge of the time since the last arrival or the time since the transmission began does not change these probabilities. The state of the queue at any time is given by the number of messages present. Denote by $\pi_n(t)$, the probability that there are n messages present at time t. Using knowledge of the properties of the Poisson process, the evolution of these probabilities at time $t + \delta t$ can be studied. Considering states with a message currently being transmitted, there are four possible occurrences in the next short interval, δt:

a) There are no arrivals, and the message being transmitted continues transmission.
b) There is an arrival, and the message being transmitted continues in transmission.
c) There are no arrivals, and the message being transmitted finishes its transmission.
d) There is an arrival, and the message being transmitted finishes its transmission.

The probability of there being more than one arrival in δt is $o(\delta t)$. Similarly, the probability that there is an arrival *and* the message currently being transmitted finishes its transmission is $\lambda \delta t \mu \delta t$, which is $o(\delta t)$. Thus if the system is in state n at time $t + \delta t$, then the system must have been in state $n - 1$, n, or $n + 1$, at time t.

$$\pi_n(t + \delta t) = \pi_n(t)\left[(1 - \lambda \delta t)(1 - \mu \delta t) + o(\delta t)\right]$$
$$+ \pi_{n+1}(t)(1 - \lambda \delta t)\mu \delta t + \pi_{n-1}(t)(1 - \mu \delta t)\lambda \delta t + o(\delta t)$$

Hence:

$$\pi_n(t + \delta t) - \pi_n(t) = (-\lambda - \mu)\delta t \pi_n(t) + \mu \delta t \pi_{n+1}(t) + \lambda \delta t \pi_{n-1}(t) + o(\delta t)$$

and dividing by δt and taking the limit as $\delta t \to 0$, the derivative of $\pi_n(t)$, $\pi_n'(t)$ is given by:

$$\pi_n'(t) = (-\lambda - \mu)\pi_n(t) + \mu \pi_{n+1}(t) + \lambda \pi_{n-1}(t) \qquad (4.1)$$

Similarly,

$$\pi_0'(t) = -\lambda \pi_0(t) + \mu \pi_1(t)$$

This system of differential equations can be solved, and the evolution of the system through time can be studied from any particular starting state. However this approach will not be followed for two reasons. First, the solution is not straightforward, and involves the infinite sum of a set of Bessel functions of the first kind. Secondly, it is known from Markov process theory that in the long run the initial state does not affect the steady state probabilities. These *steady state* or *equilibrium* probabilities give the long term behaviour of the system, and this is the object of most analyses. Since, by definition of equilibrium, the probabilities are not changing with respect to time, $\pi'_n(t)$ can be equated to 0. Writing π_n for $\lim_{t\to\infty} \pi_n(t)$, the equations (4.1) can be written as:

$$\lambda\pi_0 = \mu\pi_1 \tag{4.2}$$
$$(\lambda + \mu)\pi_1 = \lambda\pi_0 + \mu\pi_2 \tag{4.3}$$
$$(\lambda + \mu)\pi_2 = \lambda\pi_1 + \mu\pi_3 \tag{4.4}$$
$$\vdots$$

These are the global balance equations for this system. By successive elimination, they can be rewritten as:

$$\lambda\pi_0 = \mu\pi_1$$
$$\lambda\pi_1 = \mu\pi_2 \tag{4.5}$$
$$\lambda\pi_2 = \mu\pi_3$$
$$\vdots$$

Every probability can be expressed in terms of π_0.

$$\pi_1 = \frac{\lambda}{\mu}\pi_0$$
$$\pi_2 = \frac{\lambda}{\mu}\pi_1 = \frac{\lambda^2}{\mu^2}\pi_0$$
$$\vdots$$

Now λ/μ is the traffic intensity, ρ, so it follows that:

$$\pi_n = \rho^n \pi_0 \tag{4.6}$$

From a corollary to Little's theorem for any single server system, it is known that:

$$\pi_0 = 1 - \rho$$

so the steady state probabilities can be calculated for all n, $\pi_n = \rho^n(1 - \rho)$. In this case π_0 is known independently, but it is more common for an analysis to solve the infinite set of balance equations in terms of a finite set of state probabilities, which must then be found to give the probability values. The above analysis, as far as equation (4.6) is typical, expressing all probabilities in terms of π_0.

From the axioms of probability, $\sum_{n=0}^{\infty} \pi_n = 1$. This is sometimes referred to as the normalisation condition. Hence, using equation (4.6),

$$\sum_{i=0}^{\infty} \rho^i \pi_0 = 1$$

Taking out the common factor π_0 on the left hand side, we get:

$$\pi_0 \sum_{i=0}^{\infty} \rho^i = 1$$

which can be summed as a geometric series, provided that ρ is less than 1, to give:

$$\pi_0 \cdot \frac{1}{1-\rho} = 1$$

and $\pi_0 = 1 - \rho$ as expected. Notice that the condition for the stability of the queue (and the existence of a steady state distribution) appears naturally in the solution as the condition for the existence of a normalised solution to the balance equations.

Thus the number of messages in an $M/M/1$ queue is geometrically distributed with parameter ρ. In order to calculate the mean number of messages waiting the generating function of the number of messages is found.

$$\begin{aligned} G(z) &= \sum_{n=0}^{\infty} \pi_n z^n \\ &= \frac{1-\rho}{1-\rho z} \end{aligned}$$

and the mean number in the system is:

$$G'(1) = \frac{\rho}{1-\rho} \tag{4.7}$$

In this case the state probabilities were found directly, and the generating function only used to facilitate the finding of the mean. In many cases, it is difficult to find the probabilities explicitly and the problem can be solved by finding the generating function directly.

It is also worth pointing out that even in this simple example a conflict of interest is evident. Channel or server idle time is often seen by managers as an available resource, and policies are modified to reduce this idle time. Users of a service, on the other hand, see delay as the major measure of effectiveness. The utilisation of the transmission channel is given by ρ. As this utilisation increases, and approaches one, giving less and less idle time, the mean queue length increases with correspondingly longer mean waiting times. In the limit, maximum throughput can be achieved by allowing the rate of arrivals to approach the service rate, but only at the expense of allowing the mean waiting time to become unbounded.

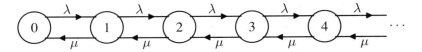

Figure 4.1 Steady state diagram for $M/M/1$ queue

4.2   Steady state diagrams

The steady state, or global balance, equations were found using the properties of the Poisson and exponential distributions to construct a set of time dependent differential equation giving the probability distribution. By equating the rate of change to zero, the steady state balance equations were found. In Markovian systems the global balance equations can easily be derived from the balance diagram. The states of the system are represented as nodes, linked by directed arcs. The arcs are labelled with the exponential (or Poisson) rate of the corresponding transition. Figure 4.1 is the steady state diagram for the $M/M/1$ queue. If the system reaches equilibrium, then for any set of states, the net 'flow' of probability into the set of states, as measured by the product of the label on the arcs and the probabilities of their starting states, must equal zero. If this were not so, then the probabilities would be changing, violating the assumption of steady state. The flow across *any* closed curve, or *cut*, must total zero. Using cuts on the balance diagram is a powerful graphical technique for deriving the balance equations. In the $M/M/1$ queue, drawing a cut surrounding each state individually leads to the first set of balance equations (4.3). However, if we use a cut which surrounds all the states, 0, 1, 2, $\ldots$, i, we derive directly the second form of the balance equations (4.5). Examination of the steady state diagram and judicious choice of cuts often leads to equivalent formulations of the balance equations which are simpler to solve.

4.3   Birth-and-death processes

The $M/M/1$ queue is an example of an infinite state *birth-and-death process*. The system population increases by at most one, and decreases by at most one at any transition. Expressed as a Markov process, the transition rate matrix is tridiagonal. We now give conditions under which an arbitrary birth-and-death process will have a steady state solution.

Consider an infinite state birth-and-death process in which the birth rate is λ_i when the system is in state i. Similarly, denote the death rate by μ_i. Assuming that a steady state exists:

$$\lambda_0 \pi_0 = \mu_1 \pi_1$$

$$\lambda_1 \pi_1 = \mu_2 \pi_2$$
$$\lambda_2 \pi_2 = \mu_3 \pi_3$$
$$\vdots$$

As before, every probability can be expressed in terms of π_0.

$$\pi_1 = \frac{\lambda_0}{\mu_1} \pi_0$$

$$\pi_2 = \frac{\lambda_1}{\mu_2} \pi_1 = \frac{\lambda_0 \lambda_1}{\mu_1 \mu_2} \pi_0$$

$$\vdots$$

In general,

$$\pi_i = \frac{\prod_{j=0}^{i-1} \lambda_j}{\prod_{j=1}^{i} \mu_j} \cdot \pi_0$$

The normalisation condition demands that $\sum \pi_i = 1$, so

$$\left(\sum_{i=0}^{\infty} \frac{\prod_{j=0}^{i-1} \lambda_j}{\prod_{j=1}^{i} \mu_j} \right) \cdot \pi_0 = 1$$

The existence of a steady state solution depends only on the convergence of the sum. If the sum converges, the steady state exists, and π_0 can be found using the normalisation condition. If the sum diverges, no steady state exists.

4.4   Limited waiting room

The next example is the $M/M/1/N$ queue, that is, the queue with Poisson arrivals, exponential service times, a single server, and only enough room for N customers to be in the system at once. This limited waiting room might represent a set of buffers allocated to one output channel in a packet switching computer providing a datagram service; an arriving packet which finds all buffers in use is simply discarded. The designer wishes to choose the number of buffers that are available to that circuit to provide an acceptably low probability that packets will be discarded. The throughput of the system, that is the rate at which messages can be transmitted by the system, is a measure of the effectiveness of the buffers. More buffers will mean fewer jobs being rejected, and should increase the throughput. In contrast, buffer space is not free, so there is an argument for reducing the value of N to the minimum acceptable level. The cost of extra buffers is known; the cost of too few buffers can be measured by calculating the rate at which messages are refused entry to the system because there are insufficient buffers.

In order to evaluate these factors, the steady state distribution of the number of messages in the system is needed. Since the processes involved are memoryless, the state of the system is given by the number of messages. The state can take any value between 0 and N. Transitions between states are the same as in the $M/M/1$ case except that when the system is in state N, there are no arrivals. The balance equations are:

$$\mu \pi_1 = \lambda \pi_0$$
$$\vdots$$
$$\mu \pi_N = \lambda \pi_{N-1}$$

which are easily solved to give:

$$\pi_i = \rho^i \pi_0, \quad \text{for } i = 1, 2, \ldots, N \tag{4.8}$$

but now π_0 cannot be equated to $1 - \rho$, because the mean arrival rate to the system is *not* λ, but $\lambda(1 - \pi_N)$, since messages attempting to arrive when the system is in state N are refused entry. From the normalisation condition, $\sum_{i=0}^{N} \pi_i = 1$, we can establish that:

$$\pi_0 = \frac{1 - \rho}{1 - \rho^{N+1}}$$

The rate at which messages arrive, but are refused admission to the system, is given by $\lambda \pi_N$. Notice that because this is a finite system, no restrictions on the values of λ and μ are necessary to ensure that equilibrium is reached. If λ is greater than μ, that is, messages arrive faster than the system can serve them, a steady state is still reached, but state N has a higher probability than state 0, i.e., the system is more likely to be full than empty. When λ and μ are equal, the probabilities of all states are identical, $\pi_i = \frac{1}{N+1}$.

The throughput of the system can be found in two equivalent ways. Either it is the rate at which jobs are accepted into the system, which equals $\lambda(1 - \pi_N)$, or it is the rate of leaving the system, which equals $\mu(1 - \pi_0)$. In a stable system the rate of arrivals will equal the rate of departures. Here it is easy to calculate both these quantities, and confirm the equality, but often one or other will be more difficult to find. The throughput of the system is an increasing function of the waiting room N; if $\lambda < \mu$ then as $N \to \infty$ the throughput tends to λ; if $\mu < \lambda$ then it tends to μ as $N \to \infty$.

The utilisation of the waiting room is given by $\frac{E[n]}{N}$. It is interesting to observe that if $\lambda = \mu$, then the utilisation is *always* $\frac{1}{2}$, regardless of the value of N. The optimum value of N can be found by calculating the throughput for successive values of N, until the increase in revenue achieved by adding a further space in the waiting room is less than the cost of adding the extra space.

4.5   Finite customer population

A Markovian queue with a finite customer population can be used to model a conventional time-sharing system. Each terminal is represented by a customer. The terminal thinks for a period which is exponentially distributed with mean $1/\lambda$, and then submits a job to the central computer which has an exponentially distributed service requirement, mean $1/\mu$. Jobs at the central computer are served in FCFS order. When the service requirement has been fulfilled, the terminal enters think state again. This is an $M/M/1//N$ queueing system.

Since all the processes are memoryless, the state of the system can be given by the number of jobs in the CPU queue. If there are N terminals, then the state can be $0, 1, 2, \ldots, N$. When there are i jobs in the CPU queue, there are $N - i$ terminals in think state, so that new jobs join the queue at rate $(N - i)\lambda$. Jobs leave the queue at rate μ. The balance equations are easily written down:

$$N\lambda\pi_0 = \mu\pi_1$$
$$(N-1)\lambda\pi_1 = \mu\pi_2$$
$$(N-2)\lambda\pi_2 = \mu\pi_3$$
$$\vdots$$
$$\lambda\pi_{N-1} = \mu\pi_N$$

It is straightforward to solve these equations in terms of $\rho(= \frac{\lambda}{\mu})$, and π_0, to give:

$$\pi_i = \frac{N!}{(N-i)!}\rho^i\pi_0$$

π_0 can now be found using the normalising condition.

$$\pi_0 = \left[\sum_{i=0}^{N}\frac{N!}{(N-i)!}\rho^i\right]^{-1}$$

The effect of the number of terminals on the mean response time of the system can now be investigated. The response time measures the time between a terminal submitting a job to the CPU, and the terminal entering think state again. Little's theorem is used. The average arrival rate is *not* λ, but $\Lambda = \sum_{i=0}^{N-1}\lambda(N-i)\pi_i$, and the mean number of jobs in the queue is:

$$\bar{N} = \sum_{i=1}^{N}\frac{N!}{(N-i)!}i\rho^i\pi_0$$

The term $N!\pi_0$ is common to both $\bar{N}$ and Λ, so the response time is given by:

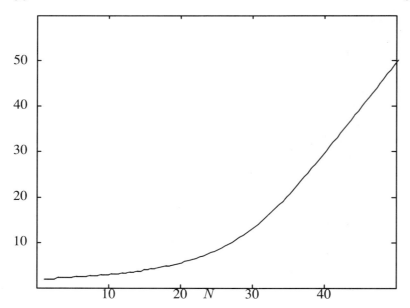

Figure 4.2 Response time of $M/M/1//N$ system

$$\bar{W} = \frac{\bar{N}}{\Lambda} = \frac{\sum_{i=1}^{N} \frac{i}{(N-i)!} \rho^i}{\lambda \sum_{i=1}^{N} \frac{1}{(N-i)!} \rho^i}$$

The effect of increasing the number of terminals on a system can now be calculated. For example, assume that users think for an average of 50 seconds, and that each request to the CPU has mean 2 seconds. ρ equals 0.04. In Figure 4.2 we plot the mean response time as a function of N, the number of users. For small N, adding extra users has only a small effect on the average response time. As N increases, the effect of extra users is to increase the response time more rapidly, until, for large values of N, each extra user is increasing the mean response time by 2 seconds.

The intuitive explanation is not hard to find. When there are a small number of users, the CPU queue will be empty for a high proportion of the time, so any additional user will absorb some of this idle time, and interfere with the other users only in as far as the users are not centrally controlled, but individual independent processes. For large values of N, the system is heavily loaded, with almost no idle time at the CPU, and each additional user only adds to the queue at the CPU, increasing the mean response time.

This model was also studied very early in the history of queueing theory. It is usually given the name *machine repairman problem*. The customers represent machines which are operational for exponentially distributed intervals, independently of one another, and then fail, and have to be repaired by a single repairman. The repair process is also exponentially distributed.

4.6   Discrete time queues

In many communications contexts, it is convenient to treat time as a discrete quantity. For example, if packets can only have one of a few different sizes, it may be useful to treat the time as discrete, with one unit of time representing the shortest possible packet's transmission time. Some simple discrete time queues will be studied. It might be hoped that they would be simpler to analyse than continuous time queues, but it appears that if anything they are more complex. In continuous time, only an arrival or a service completion could occur at one instant, not both. This is not so in discrete time systems, where simultaneous arrivals and departures must be allowed. The discrete time intervals are usually known as *slots*. It may be that only a single arrival can occur in any slot, or multiple arrivals may be possible.

Perhaps the simplest of all discrete time queues is analogous to the $M/D/1$ queue. Only single packet arrivals are allowed. The state of the system is given by the number of packets left to be transmitted. A single new packet arrives with probability α, and no new packets arrive with probability $1 - \alpha$. If there are any queued packets, one is transmitted in every slot. If the system starts off empty, then the only possible states of the system are 0 and 1. If the system is in state 0, it is empty, and it makes the transition to state 1 at the next time with probability α. When the system is in state 1, if no arrival occurs, then it will be in state 0 at the next time instant, and if an arrival occurs it will still be in state 1 at the next instant, since the arriving job will exactly balance the job which is served. Even if the system starts off with a long initial queue of packets, with probability 1 it will return to state 0, and once that state has been reached, the only other state that can be reached is state 1. This is easily seen, since in any state where there are jobs to be served, a job will be served, so the state of the system will reduce by one unless there is an arrival. Therefore any states other than 0 and 1 are transient, and have steady state probability 0. It is obvious that the steady state probabilities are $\pi_0 = 1 - \alpha$ and $\pi_1 = \alpha$. This is not a very interesting system.

The previous queue can be generalised by allowing several packets to arrive in the same time slot. This could represent a continuous arrival process which is only observed at the end of slots. Another possible motivation in the field of communication networks is the possibility that a message is made up of a number of fixed-length packets. When a new message arrives to be transmitted, several packets are added to the queue simultaneously. If the queue being analysed takes its input from the output of several other queues, then even though each queue can only transmit a single packet in a slot, it is possible for several new packets to be added to the queue in the same slot. The probability of i arrivals in a slot will be denoted by α_i, with generating function $A(z) = \sum_{i=0}^{\infty} \alpha_i z^i$. One packet will be removed from the queue in each slot. What are the balance equations of this system? It is convenient to assume that the probability of leaving any state in a slot is 1, even if the same state is reentered at the next slot. Consider state 0, arrivals to the state occur from state 0, if no arrivals occur, with probability α_0, and from state 1, also if no arrivals occur. Hence:

$$\pi_0 = \alpha_0 \pi_0 + \alpha_0 \pi_1$$

In state 1, arrivals occur from state 0 with probability α_1, from state 1 when there is a single arrival which has probability α_1, and from state 2 if there is no arrival, probability α_0. Expressed as a balance equation, we have:

$$\pi_1 = \alpha_1 \pi_0 + \alpha_1 \pi_1 + \alpha_0 \pi_2$$

In general, the arrivals to state j are from state 0 with probability α_j, and from state k for $k = 1, 2, \ldots, j+1$ with probability α_{j+1-k}, giving the balance equation:

$$\pi_j = \alpha_j \pi_0 + \sum_{k=1}^{j+1} \alpha_{j+1-k} \pi_k \tag{4.9}$$

We find the generating function of the number of messages in the system,

$$P(z) = \sum_{i=0}^{\infty} \pi_i z^i$$

by multiplying each of equations (4.9) by z^j, and summing:

$$
\begin{aligned}
P(z) &= \sum_{j=0}^{\infty} \alpha_j z^j \pi_0 + \sum_{j=0}^{\infty} z^j \sum_{k=1}^{j+1} \alpha_{j+1-k} \pi_k \\
&= A(z)\pi_0 + \sum_{k=1}^{\infty} \sum_{j=k-1}^{\infty} \alpha_{j+1-k} z^j \pi_k \\
&= A(z)\pi_0 + \sum_{k=1}^{\infty} \pi_k z^{k-1} \sum_{j=k-1}^{\infty} \alpha_{j+1-k} z^{j+1-k} \\
&= A(z)\pi_0 + A(z)\frac{P(z) - \pi_0}{z} \\
&= \pi_0 \frac{A(z)(z-1)}{z - A(z)}
\end{aligned}
$$

In order to find π_0, we use the normalisation condition, $P(1) = 1$. Direct substitution of $z = 1$ leads to an indeterminacy of the form $0/0$, which is resolved using l'Hospital's rule to give:

$$\pi_0 = 1 - A'(1)$$

which agrees with our formula in the case of only single arrivals being possible given above. (In this case, $A(z) = 1 - \alpha(1 - z)$.) Although we have not commented on it, the condition for stability of the system is clearly $A'(1) < 1$. Although we could evaluate the individual state probabilities, if needed, and the moments of the queue length from the generating function, we will generalise the system slightly, including this model as a special case.

We consider multipacket arrivals as above, but instead of the queue shortening by one packet at every time interval, it will only reduce with probability β. This might correspond to a noisy communication channel in which the probability of successful transmission of a packet was β, or to a system in which the packets had geometrically distributed lengths with mean $1/\beta$. Now the probability that there will be one less packet in the queue is not α_0, but $\alpha_0 \beta$. Similarly, the probability that the queue length is unchanged is $\alpha_1 \beta + \alpha_0 (1 - \beta)$, if there are packets in the queue, and α_0 if the system is idle. In general, the probability that the queue increases in length by $j - 1$ packets is given by $\alpha_j \beta + \alpha_{j-1}(1 - \beta)$. Clearly, our original model corresponds to the case $\beta = 1$. Writing $B(z) = \beta + (1 - \beta)z$, the equation for the generating function of queue length comes out as:

$$\begin{aligned} P(z) &= A(z)\pi_0 + A(z)B(z)\frac{P(z) - \pi_0}{z} \\ &= \pi_0 \frac{A(z)(z - B(z))}{z - A(z)B(z)} \end{aligned}$$

and we can evaluate π_0 from the normalisation condition to find:

$$\pi_0 = 1 - A'(1)/\beta$$

Again we note without further comment that for stability $A'(1) < \beta$. This corresponds to the mean number of arrivals in a slot being strictly less than the mean number of packets which can be served in a slot. In order to evaluate the mean queue length, we have to evaluate the derivative of $P(z)$ at $z = 1$. We note that indeterminacy of the form $0/0$ arises again. Again we will use l'Hospital's rule to resolve it. We follow through the working in detail here, since a number of short cuts arise. First, define:

$$P(z) \triangleq \frac{u(z)}{v(z)} \quad \text{with } u(1) = v(1) = 0$$

Hence,

$$P'(z) = \frac{v(z)u'(z) - u(z)v'(z)}{v(z)^2}$$

which is obviously of the form $0/0$ at $z = 1$. After applying l'Hospital's rule once, we have:

$$P'(z) = \frac{v(z)u''(z) - u(z)v''(z)}{2v(z)v'(z)}$$

which still evaluates to $0/0$. A second application of the rule leaves:

$$P'(z) = \frac{v'(z)u''(z) + v(z)u'''(z) - u'(z)v''(z) - u(z)v'''(z)}{2v(z)v''(z) + 2v'(z)^2}$$

Evaluation of this formula at $z = 1$ is less formidable than at first sight it appears, since $u(1) = v(1) = 0$. Simplifying,

$$P'(1) = \frac{v'(1)u''(1) - u'(1)v''(1)}{2v'(1)^2} \tag{4.10}$$

Substituting in the appropriate values, remembering that in this case $B''(1) = 0$, we find:

$$P'(1) = A'(1) + \frac{A''(1) + 2A'(1)B'(1)}{2(1 - A'(1) - B'(1))}$$

In the case where we only have a single arrival in a slot, with probability α, $A(z) = (1 - \alpha) + \alpha z$, and the mean queue length is given by:

$$P'(1) = \frac{\alpha(1 - \alpha)}{1 + \beta - \alpha}$$

4.7 Exercises

1. Compare two $M/M/1/n$ queueing systems. System (a) has $n = 2N$, and service rate μ, while system (b) has $n = N$, and service rate 2μ. Which system gives better service? Does the comparative worth change if the service rates of both systems are μ, but the arrival rates are λ and $\lambda/2$, respectively?

2. Consider the $M/M/1$ system with discouraged arrivals. Arrivals at the system, which occur in a Poisson stream at rate λ, will only join the system with probability $1/(i + 1)$ where i is the number of jobs currently in the system. With probability $i/(i + 1)$ they leave immediately. Find the steady state distribution of the number in the system, and the throughput of the system in terms of completed services per unit time.

3. Consider the $M/M/1$ queue with bulk arrivals. A job consists of one or two tasks with equal probability, and jobs arrive at rate λ. Draw the steady state diagram, and hence, or otherwise, find the steady state equations. Solve the equations to find the probability distribution of the number of *tasks* in the system. Under what conditions is a steady state reached?

4. Extend the analysis of the previous exercise to the general bulk arrival Markovian queue. That is, assume that jobs arrive in batches of size j with probability p_j. The inter batch time is exponentially distributed, as is the service time of each job.

4.8 Further reading

The $M/M/1$ queue has been extensively studied. A.K. Erlang, the father of queueing theory, used variations on $M/M/1$ queues in his pioneering study of telephone exchanges[1]. Most authors, including Saaty[5] and Kleinrock[2], give extensive coverage to $M/M/1$ queues. Both these books develop the time dependent solution. Scherr[6] used the $M/M/1//N$ queueing system developed in section 4.5 to analyse the CTSS system at MIT. This pioneering application of queueing theory to computer systems was the inspiration for much work on analysis of computer system performance. Despite its simplicity, the Markovian queue is still the object of research. Pegden was able to write a paper with 'new results' in 1982[4].

As we observed, discrete time queues are significantly more complex than their continuous time equivalents. Meisling[3] is a simple introduction to the field.

4.9 Bibliography

[1] A.K. Erlang. Probability and telephone calls. *Nyt Tidsskr. Mat., Ser. B*, 20:33–39, 1909.

[2] L. Kleinrock. *Queueing Systems Volume 1: Theory*. Wiley, New York, NY, 1975.

[3] T. Meisling. Discrete-time queueing theory. *Operations Research*, 6(1):96–105, January/February 1958.

[4] C.D. Pegden. Some new results for the $M/M/1$ queue. *Management Science*, 28(7):821–828, 1982.

[5] T.L. Saaty. *Elements of Queueing Theory*. McGraw-Hill, New York, NY, 1961.

[6] A.L. Scherr. *An Analysis of Time-Shared Computer Systems*. MIT Press, Cambridge, Massachusetts, 1967.

Chapter 5

$M/G/1$ **queues**

Markovian queues such as were studied in the previous chapter can be analysed very simply, but many systems have more complex behaviour. For example, in a packet switching system, the packets are not exponentially distributed in length, but are either constant length, or of a random length with some minimum and maximum lengths imposed by the protocols of the system. In fact, measurements on the packet lengths in real packet switching networks show that the distribution of packet lengths is bi-modal; most of the packets are either short, corresponding to interactive terminal traffic, or long corresponding to file transfers and other high volume transactions.

This chapter is concerned with the simplest of these more general systems: the $M/G/1$ queue. Messages with generally distributed lengths arrive to be transmitted on a single channel. The arrivals form a Poisson stream. The number of messages no longer determines the future behaviour of the system by itself. The probability that a transmission in progress will end within a particular time is no longer independent of time, but depends on how long the transmission has lasted already. Hence the state of the system can be given by the number of messages present *and* the remaining transmission time of the message currently being transmitted (if any).

There are many approaches to the analysis of this system; we shall examine several. The first technique treats the $M/G/1$ queue only. The other two can be adapted to other circumstances, and are important parts of a performance analyst's toolbox of techniques. No assumptions are made about the distribution of the length of each message, except that the first and second moments should be finite. Message lengths are independent of arrival times, and of the lengths of the other messages. Since transmissions occur at a fixed rate, message length and transmission times are directly related. We denote the distribution function of transmission times by $B(t)$, its mean by $E[S]$ and its second moment by $E[S^2]$. Arrivals are independent of each other and of the message lengths. The inter-arrival time is exponentially distributed with parameter λ.

5.1 Mean queue length

We consider the length of the queue at the instants just after a message departs, but before the next message in the queue has started transmission. These instants are not evenly spaced in time, since messages have different service requirements. If the departure left the system empty the interval until the next departure will include an idle period, until a new message arrives, as well as the transmission time of a message. Let us denote the number of messages left behind by the ith message on departure by n_i. Similarly denote the number of arrivals during the transmission of the ith message as a_i. How does n_{i+1} depend on these quantities? The new queue length is the old queue length, less the message which has just finished transmission, plus any arrivals that occurred during the transmission of the message which has just departed. If the queue was empty after the last departure, the new queue length will be just those messages that arrived during the transmission of the job which has just departed.

$$n_{i+1} = \begin{cases} n_i - 1 + a_{i+1}, & n_i > 0 \\ a_{i+1}, & n_i = 0 \end{cases} \tag{5.1}$$

Defining the function $\delta(x)^\dagger$ by:

$$\delta(x) = \begin{cases} 1, & x > 0 \\ 0, & x = 0 \end{cases}$$

we can rewrite the formula as:

$$n_{i+1} = n_i - \delta(n_i) + a_{i+1} \tag{5.2}$$

We take the expectations of both sides, and get:

$$E[n_{i+1}] = E[n_i] - E[\delta(n_i)] + E[a_{i+1}] \tag{5.3}$$

So far, none of our operations has depended on the distribution of the quantities involved, although we have implicitly assumed that the expectations exist and are finite. Since, as always, we are seeking an equilibrium solution, the expectations must be stationary, and independent of the subscript. Therefore:

$$E[n_{i+1}] = E[n_i] = E[n]$$

The assumption of Poisson arrivals means that $E[a_{i+1}]$ will be independent of i, say $E[a]$. If arrivals were not Poisson, then the number of arrivals during the transmission of the ith message would depend on when the $i-1$st message finished transmission with respect to the arrival process. We now rearrange equation (5.3) and find:

$$E[\delta(n)] = E[a]$$

---

†This $\delta(x)$ function is sometimes known as the Kronecker delta function.

Since arrivals are Poisson with rate λ, the mean number of arrivals in an interval of length t is λt. Hence, the expected number of arrivals in a service interval is $\int_0^\infty \lambda t \, dB(t) = \lambda E[S] \stackrel{\triangle}{=} \rho$.

That is, the probability that the queue left by a message departing after service is non-zero is equal to the expected number of arrivals during a message's transmission. This is just a restatement of the result that we derived in the first chapter for the server idle probability in a $G/G/1$ queue.

To make any further progress, we examine equation (5.2) again. It is an identity, so if we square both sides, they will still be equal. We first observe that for non-negative x, $\delta(x)^2 = \delta(x)$ and $x \cdot \delta(x) = x$. Squaring, we arrive at:

$$n_{i+1}^2 = n_i^2 + \delta(n_i) + a_{i+1}^2 - 2n_i - 2\delta(n_i)a_{i+1} + 2n_i a_{i+1}$$

and taking expectations of both sides,

$$E[n^2] = E[n^2] + E[\delta(n)] + E[a^2] - 2E[n] - 2E[\delta(n)a] + 2E[na] \tag{5.4}$$

Since arrivals are independent of queue length,

$$E[\delta(n)a] = E[\delta(n)]E[a]$$

and:

$$E[na] = E[n]E[a]$$

Rearranging (5.4) we immediately find that:

$$E[n] = E[a] + \frac{E[a^2] - E[a]}{2(1 - E[a])} \tag{5.5}$$

We have already calculated $E[a]$; it equals ρ. $E[a^2]$ is the only remaining unknown. If we assume that the transmission time of the message just departed was t, then the probability of i arrivals is $(\lambda t)^i / i! \cdot e^{-\lambda t}$, since arrivals are Poisson, and $E[a^2|t] = \lambda t(1 + \lambda t)$. To remove the conditioning on the value of t, we use the continuous version of the theorem of total probability, integrate with respect to $B(t)$, the distribution function of service time, and find:

$$E[a^2] = \int_0^\infty \lambda^2 t^2 + \lambda t \, dB(t)$$
$$= \lambda^2 E[S^2] + \lambda E[S]$$

Substituting in (5.5), we arrive at the Pollaczek-Khintchine formula:

$$E[n] = \rho + \frac{\lambda^2 E[S^2]}{2(1 - \rho)} \tag{5.6}$$

This is sometimes expressed in terms of the squared coefficient of variation of the service time, C_b^2, giving:

$$E[n] = \rho + \frac{\rho^2(1 + C_b^2)}{2(1 - \rho)} \qquad (5.7)$$

The exponential distribution has squared coefficient of variation equal to 1, and on substituting into this formula we recover our expression for the mean of an $M/M/1$ queue (4.7). The formula for the $M/M/1$ queue gave the steady state mean queue length at any time. The formula just derived gives the average number in the queue at departure instants. It is possible to verify that this is identical to the mean queue length at *any* instant. The implications of the formula are surprising, at least initially. The mean queue length, and because of Little's theorem the mean waiting time, depend not only on the mean message length, but also on the variance of message lengths. Notice that if the variance of the service time is infinite, then so is the mean queue length, even though all the distributions may have finite means and the system is not saturated. If two streams of messages have the same mean length, but different variances, then they will experience different mean waiting times. More specifically, if one stream is composed of messages of constant length l, and the other stream is composed of messages of length $l/2$ and $3l/2$, each length of message occurring with probability 0.5, then although they have the same mean message length, the mean queue length and mean waiting time will be greater in the system with two lengths of message. The formula (5.7) shows that the minimum queue length will be obtained when the variance of message lengths is reduced to 0. This is a general phenomenon; the queue lengths and waiting times in a system will always be reduced if the variance of the random processes can be reduced.

5.2   Embedded Markov chain

In the previous section, we observed, without commenting on it, that the number of messages left by a departing message depended on the number left by the previous departure *and* on the number of arrivals during the service. Since arrivals are Poisson, and we know the distribution of message lengths, we can calculate the distribution of the number that will arrive during a service. Thus we can calculate the probability of the different state changes, and the probability of the number of messages left by a departing message having a particular value depends only on the number of messages left by the previous departure, and not on its history. That is, the queue lengths at departure instants form a Markov chain.

As before we consider the system at departure instants only. The state of the system is given by the number of messages in the queue, and we denote the steady state probability of there being j messages by π_j. We need to find the transition probabilities, p_{ij}, from state i to state j. When these transition probabilities are known, π_i are the solutions to the balance equations:

$$\pi_j = \sum_{i=0}^{\infty} p_{ij}\pi_i \tag{5.8}$$

We have already observed that the transition probabilities depend only on the number of arrivals during a transmission. This is reflected in equation (5.1). Let a_i be the probability of i arrivals during a message's transmission. The discussion used above when we evaluated the mean queue length applies here, and shows that $p_{ij} = a_{j-i+1}$ for $i > 0$, and $p_{0j} = a_j$. Transitions from i to j with $j < i - 1$ are impossible, since they imply more than one departure from the queue. Hence (5.8) becomes:

$$\pi_j = \sum_{i=1}^{\infty} a_{j-i+1}\pi_i + a_j\pi_0 \tag{5.9}$$

A common tool for dealing with infinite sets of equations such as these is the generating function. Let $A(z) = \sum_{i=0}^{\infty} a_i z^i$ be generating function of the number of arrivals during a message's transmission, and $P(z)$ be the generating function of the queue length left by a departing job. We multiply each of (5.9) by z^j, and add them.

$$\sum_{j=0}^{\infty} \pi_j z^j = \sum_{j=0}^{\infty}\sum_{i=1}^{\infty} a_{j-i+1}\pi_i z^j + \sum_{j=0}^{\infty} a_j\pi_0 z^j$$

Rearranging the sums, we find:

$$P(z) = \sum_{j=0}^{\infty} \pi_j z^j = \sum_{i=1}^{\infty} \pi_i z^{i-1} \sum_{j=i-1}^{\infty} a_{j-i+1}z^{j-i+1} + A(z)\pi_0$$

which simplifies to:

$$P(z) = A(z)\frac{P(z) - \pi_0}{z} + A(z)\pi_0$$

Hence:

$$P(z) = \frac{A(z)(1 - z)}{A(z) - z}\pi_0$$

Now we can evaluate $A(z)$ simply. Given that a transmission lasts t time units, the probability of i arrivals, a_i, is given by $\frac{(\lambda t)^i}{i!}e^{-\lambda t}$. Thus the generating function of the number of arrivals during a transmission, conditional on the transmission having duration t is given by:

$$A_t(z) = \sum_{i=0}^{\infty} \frac{(\lambda t)^i}{i!}e^{-\lambda t} \cdot z^i$$

$$= e^{-\lambda t(1-z)}$$

Removing the conditioning on t, we find that:

$$
\begin{aligned}
A(z) &= \int_0^\infty A_t(z)\,\mathrm{d}B(t)\\
&= B^*(\lambda(1-z))
\end{aligned}
$$

the Laplace transform of the transmission time, evaluated at $\lambda(1-z)$. Hence:

$$
P(z) = \frac{B^*(\lambda(1-z))(1-z)}{B^*(\lambda(1-z)) - z}\pi_0 \tag{5.10}
$$

Although we already know that π_0 is $1 - \rho$, from the corollary to Little's result, it is instructive to verify this using the normalisation condition $P(1) = 1$. Direct substitution of $z = 1$ leads to an indeterminate form $0/0$. Using l'Hospital's rule, we find:

$$
-\lambda B^{*\prime}(0) - 1 = \pi_0 \left[(1-z)(-\lambda B^{*\prime}(0)) - B^*(0) \right]
$$

Now $-B^{*\prime}(0) = E[S]$, and $B^*(0) = 1$, so $\pi_0 = 1 - \rho$, as expected.

We can readily confirm the formula arrived at in the previous section for the mean queue length, by differentiating $P(z)$ and evaluating it at $z = 1$.

5.2.1   Khintchine's argument

The analysis in the previous section gives the generating function of queue length, as observed just after the termination of service of a job. It also gives the steady state queue length distribution as observed by a random observer. In general these may be different. Khintchine proved that in a queueing system with Poisson arrivals, not in batches, and in which customers leave singly, the steady state distribution of queue length is identical to the distribution embedded at departure (or arrival) instants.

Heuristically, this is so because the number of transitions from state j to state $j + 1$ in a long period must equal the number of transitions from state $j + 1$ to state j. A system with bulk arrivals or service in bulk would not have this property.

5.3   Tagged jobs

Another powerful technique which we shall demonstrate on the $M/G/1$ queue is the *tagged job* methodology. We examine the system as seen by a particular job, chosen at random, and develop our performance measures conditioned on the system states encountered by that job. These conditional performance measures are converted to unconditional measures by using the theorem of total probability.

Consider an $M/G/1$ system, as before, with FCFS service discipline. Let the mean number of customers in the system be N and the mean number of customers

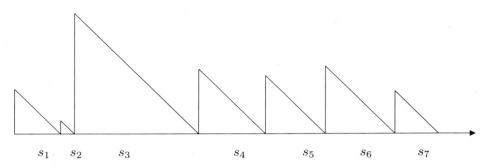

Figure 5.1 Residual service time

queueing, *not* including the customer in service, be Q. Similarly, let the mean waiting time be W and the mean queueing time be V.

Let our tagged job arrive to find n jobs waiting in the queue, with service requirements $s_1, s_2, \ldots, s_n$. The tagged job's queueing time is made up of the time until the job currently being served, if any, departs, plus the sum of the service times of all the jobs currently in the queue. If we denote the time required to complete the current job's service by r, then the queueing time of the tagged job is given by v,

$$v = \delta(n)r + \sum_{i=1}^{n} s_i$$

Take expectations of both sides, observing that the number of customers queueing is independent of the length their service requests[‡], and it follows that:

$$V = (1 - \pi_0)R + Q \cdot E[S]$$

Using Little's theorem, $Q = \lambda V$, and also $\pi_0 = 1 - \rho$, so we find:

$$V = \frac{\rho R}{1 - \rho}$$

where R is the expected time to the next departure, given that a job is currently being served. This is the *residual life* of the service time distribution. We now need to calculate its value. Figure 5.1 shows a graph of the first seven residual service times during an interval during which there were m arrivals. The idle periods have been removed since we have already conditioned the probability on there being a job in service. It is readily seen that the total area under the graph is $\sum_{i=1}^{m} \frac{1}{2} s_i^2$, and the total length is $\sum_{i=1}^{m} s_i$. Hence the expected value of the residual service time

---

[‡]This is true for any service discipline that makes no use of the length of service requested to choose which job to serve next.

observed by a random arrival in this interval is $\sum s_i^2/(2\sum s_i)$. This is just the ratio of the average area of one of the triangles to the average service time observed in the interval. As we consider longer and longer intervals, these averages will approach the expectations of the functions involved. The expected values are $\int_0^\infty x^2\,\mathrm{d}B(x)$ and $\int_0^\infty x\,\mathrm{d}B(x)$.

Thus:

$$R = \frac{1}{2}\frac{E[S^2]}{E[S]} \tag{5.11}$$

and hence:

$$V = \frac{\rho E[S^2]}{2E[S](1-\rho)} = \frac{\rho E[S](1+C_b^2)}{2(1-\rho)} \tag{5.12}$$

Now:

$$W = E[S] + V$$

and so applying Little's theorem,

$$N = \lambda W = \rho + \lambda V = \rho + \frac{\rho^2(1+C_b^2)}{2(1-\rho)}$$

again confirming the Pollaczek-Khintchine formula found in previous sections.

5.4   Random observations

The previous section shows that a random observation does not necessarily observe an arbitrary job with the same probability that an observer of the complete system evolution through time will observe. The random observation is biased towards observing the longer intervals. If the random arrival in the last section observed a random customer in service, then the expected residual service time would be half the expected service time. The fact that this is not so can be justified by observing Figure 5.1. A random arrival is equally likely to fall in any *fixed length* short interval, but there are more short intervals in the long service times than in the short service times; a random observer is more likely to observe during a long service than during a short one.

Consider a series of events, $\varepsilon_1, \varepsilon_2, \ldots$, which occur at times $t_1, t_2, \ldots$, as shown in Figure 5.2. They have a known inter-event time distribution, $F(x) = \Pr(t_{k+1} - t_k < x)$, and probability density $f(x) = \frac{dF}{dx}$. Let us now derive the distribution function of the time from a random observation until the next event. This is known as the *random modification* of the inter-event distribution, F, or as the *residual life*. Let X stand for the observed inter-event interval, and Y for the residual life.

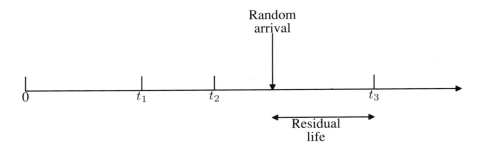

Figure 5.2 Residual life

Let the residual life have distribution function $R(x)$ and probability density function $r(x)$. Let $F_R(x)$ and $f_R(x)$ be the distribution and density functions associated with *observed* intervals. Consider $f_R(x)dx = \Pr(x < X \leq x + dx)$. We expect $f_R(x)$ to depend on $f(x)$. If long intervals are rare, then they will only be observed rarely. However our observation above about random observations being more likely to fall in a long interval than a short one suggests that $f_R(x)$ should depend on x as well. This suggests that $f_R(x) \propto x f(x)$. Since $f_R(x)$ must satisfy the constraints for a probability density function,

$$\int_0^\infty f_R(x) \, \mathrm{d}x = \int_0^\infty K x f(x) \, \mathrm{d}x = 1$$

Thus the constant of proportionality K, must equal $1/\int_0^\infty x f(x)\,\mathrm{d}x$, which is 1 over the mean interval length. Denoting the moments of the interval distribution by m_i,

$$m_i = \int_0^\infty x^i f(x) \, \mathrm{d}x$$

we find:

$$f_R(x) = \frac{x f(x)}{m_1}$$

and:

$$F_R(x) = \int_0^x f_R(x) \, \mathrm{d}x$$

Now consider our random observer, arriving in a service interval of length x. The observation of the residual service time is certainly less than x, and must be greater than 0, and there is no reason to suppose any distribution other than the uniform one between these limits, so:

$$R(y|X = x) = \Pr(Y \leq y|X = x) = \frac{y}{x}$$

This is the distribution of the residual life, *conditioned* on the length of the service interval that is *observed*. The density function is constant, $1/x$. In order to find the unconditional density of the residual life, we must remove the conditioning. Clearly a residual life of y can only occur if the service interval is longer than y, so the lower limit on the integration is y, and:

$$
\begin{aligned}
r(y) &= \int_{x=y}^{\infty} \frac{1}{x} f_R(x) \, dx \\
&= \int_{y}^{\infty} \frac{1}{x} \frac{x f(x)}{m_1} \, dx \\
&= \frac{1}{m_1} \int_{y}^{\infty} f(x) \, dx \\
&= \frac{1}{m_1} (1 - F(y))
\end{aligned}
$$

The distribution function of the residual life is given by the integral of $r(y)$, and the Laplace transform, $R^*(y)$, by:

$$
\begin{aligned}
R^*(s) &= \int_{0}^{\infty} e^{-sy} r(y) \, dy \\
&= \frac{1 - F^*(s)}{s m_1}
\end{aligned}
\tag{5.13}
$$

5.5 $M/G/1$ queues with server vacations

The analysis given in previous sections assumes that the server is immediately available to start serving new jobs which arrive when the server is idle. This is not always the case. Many communication systems work with a central station *polling* its subordinates. In some order, each subordinate station is offered the chance to transmit some or all of its messages. If it has no messages to send, then the subordinate station waits until the next time it is polled. From the point of view of a subordinate station, the server takes vacations. If it is polled, and found to be empty, then the server is unavailable for a vacation interval thereafter. We analyse a system with the added feature that after a successful service, the server is unavailable for a relaxation interval, before the queue is again examined. This relaxation interval may be 0, if the server vacations are the only extra absences of the server, or it may be non-zero, representing time taken to reset the server to a usable state. Such servers are sometimes known as *walking servers*.

As before, arrivals are Poisson, at rate λ, and services are of length S, with distribution function, $B(t)$. Assume that vacation periods are of length $\hat{S}$, independently drawn from distribution function, $C(t)$. The Laplace transform of $B(t)$

is $B^*(s)$, and of $C(t)$ is $C^*(s)$. After a service finishes, the server has a relaxation period before examining the queue again. This relaxation period is also represented by a random variable, Y. The sum of the service time and the relaxation period has Laplace transform $D^*(s)$. After the relaxation period, or after a vacation comes to an end, the queue is examined, and either a new service is started or a new vacation starts. Call the instants at which this examination of the queue takes place *test instants*. Consider the embedded chain defined by the queue length at these test instants. The state transitions of this embedded chain depend only on the queue length, as they did on the $M/G/1$ case without vacations. If the system is now empty, then it must either have been empty at the last test instant, or there was a single job present at the last test instant. In either case, no arrivals have occurred. If it was empty at the last test instant, then a vacation was taken, and during that vacation no arrivals have occurred. Alternatively, if there was a single job in the system, then that job was served and a relaxation period taken, and there were no arrivals during the service or relaxation periods, then the system will now be empty. Writing $\hat{a}_j$ for the probability of j arrivals during a vacation, and $\tilde{a}_j$ for the probability of j arrivals during a combined service and relaxation period, and π_i for the steady state probability that the queue length at a test instant is i:

$$\pi_0 = \pi_0 \hat{a}_0 + \pi_1 \tilde{a}_0$$

Similarly, we find that:

$$\pi_k = \pi_0 \hat{a}_k + \sum_{i=1}^{k+1} \pi_i \tilde{a}_{k+1-i}$$

Using $\hat{A}(z)$ for the generating function of the number of arrivals during a vacation, and $\tilde{A}(z)$ for the generating function of $\tilde{a}_j$, similar analysis to that used for the ordinary $M/G/1$ queue gives the generating function of the queue length, $P(z) = \sum \pi_k z^k$, as:

$$P(z) = \pi_0 \frac{\tilde{A}(z) - z\hat{A}(z)}{\tilde{A}(z) - z}$$

The number of arrivals is Poisson distributed, so the generating function of the number arriving during an interval of length t is $e^{-\lambda t(1-z)}$. Since the distribution of the time intervals involved in the $\tilde{a}_j$ arrivals is that of a service plus relaxation interval, $\tilde{A}(z) = D^*(\lambda(1-z))$. Similarly, $\hat{A}(z) = C^*(\lambda(1-z))$. Hence,

$$P(z) = \pi_0 \frac{D^*(\lambda(1-z)) - zC^*(\lambda(1-z))}{D^*(\lambda(1-z)) - z}$$

Using the fact that $P(1) = 1$, and l'Hospital's rule,

$$\pi_0 = \frac{1 - \lambda E[S + Y]}{1 - \lambda E[S + Y] + \lambda E[\hat{S}]}$$

This is *not* the stationary queue length distribution, or the probability that a random observer will find an empty queue; it relates to the probabilities only at the test instants. Khintchine's argument implies that the stationary distribution is that which applies just before an arrival, or just after a departure, in a single server system with no bulk arrivals. The stationary distribution can be derived from this distribution at test instants. Consider the system at those instants when a service has just terminated, and denote the probability that the queue length at these instants is n by p_n. How is p_n related to the state probabilities at the last test instant? Any arrivals since then have been added to the queue, and a single job has left. In fact, the last test instant could *not* have encountered an empty queue or there would not have been a service to terminate! The time since the last test instant is distributed as a service interval, and the probability of i arrivals is a_i. Since the distribution at the last test instant is conditional on there having been a non empty queue, we have:

$$p_n = \sum_{i=1}^{n+1} \frac{\pi_i}{1 - \pi_0} a_{n+1-i}$$

Following the now familiar path, define the generating function of p_j as $N(z)$, and:

$$N(z) = \frac{(P(z) - \pi_0)B^*(\lambda(1 - z))}{z(1 - \pi_0)}$$

This simplifies to:

$$N(z) = \frac{1 - \lambda E[S + Y]}{\lambda E[\hat{S}]} \cdot \frac{B^*(\lambda(1 - z))(1 - C^*(\lambda(1 - z)))}{D^*(\lambda(1 - z)) - z}$$

Taking the derivative of this will give the mean queue length as:

$$N = \rho + \frac{\lambda^2 E[(S + Y)^2]}{2(1 - \lambda E[S + Y])} + \frac{\lambda E[\hat{S}^2]}{2E[\hat{S}]} \tag{5.14}$$

where $\rho = \lambda E[S]$. If there is no relaxation period, $Y = 0$, and we find that the mean queue length is the Pollaczek-Khintchine formula, *plus* a term representing the arrivals during vacations. If $\hat{S}$ is also zero, then the formula reduces to the Pollaczek-Khintchine formula.

5.5.1 Disk sector queueing

Fixed head disks, or drums, are popular devices to use for backing store for paging systems. The performance of the paging system depends on the response time to request for pages from the backing store. In a fixed head disk, the read/write heads do not have to move between different cylinders, but are fixed, with one head per cylinder. A single cylinder is selected for a read or write operations and the appropriate head is activated. The time to switch between cylinders is negligible. Paging systems deal with fixed size blocks of storage, and we can assume that each cylinder can hold N blocks. Each block occupies one sector of the disk.

Assume that the disk rotates in τ seconds, and that successive request are uncorrelated, and uniformly distributed between sectors. What will be the response time to requests? The transfer of one block takes a constant τ/N seconds, since that is the time that a single sector takes to pass under the heads. If the requests are organised as a FCFS queue, then the service time for a request can be as small as τ/N if the required sector is just about to pass under the heads, or as large as $\tau(1 + 1/N)$, if the start of the appropriate sector has just passed the heads. Since it is the time for the start of the sector to come under the heads that determines the transfer time, the service time for a request arriving to find the system empty is uniformly distributed between those limits. Arrivals that find the system busy automatically become eligible for service at a sector boundary, so their service times can only be $i\tau/N$, for $i = 1, 2, \ldots, N$. Using the result of exercise 2, the generating function of the number in the system is:

$$P(z) = \pi_0 \frac{B_0^*(\lambda(1 - z))z - B^*(\lambda(1 - z))}{z - B^*(\lambda(1 - z))}$$

where $B_0^*(s)$ is the Laplace transform of the service time for the job which finds an idle system, and $B^*(s)$ the Laplace transform of the service time distribution for other jobs. π_0 is given by the normalisation condition:

$$\pi_0 = \frac{1 - \lambda\tau(N + 1)/(2N)}{1 + \lambda\tau/(2N)}$$

The mean queue length can also be calculated, and is given by

$$Q = \frac{\lambda E[S_0]}{1 - \lambda(E[S] - E[S_0])} + \frac{\lambda^2}{2} \cdot \frac{E[S_0^2] - E[S^2]}{1 - \lambda(E[S] - E[S_0])} + \frac{\lambda^2}{2} \cdot \frac{E[S^2]}{1 - \lambda E[S]}$$

with S_0 the exceptional first service time, and S the ordinary service time. Simple arithmetic shows that

$$E[S] = \frac{\tau(N + 1)}{2N}, \qquad E[S^2] = \frac{\tau^2}{6N^2}(2N^2 + 3N + 1)$$

and

$$E[S_0] = \frac{\tau(N + 2)}{2N}, \qquad E[S_0^2] = \frac{\tau^2}{3N^2}(N^2 + 3N + 3)$$

Hence the mean delay to requests to the disk is

$$W = \frac{\tau(N+1)}{2N + \lambda\tau} + \frac{\lambda}{2}\frac{\tau^2(3N+5)}{3N(2N+\lambda\tau)} + \frac{\lambda}{2}\frac{\tau^2(2N^2+3N+1)}{3N(2N-\lambda\tau(N+1))}$$

However suppose that instead of forming a single queue of requests, the system maintains separate queues for each of the N sectors. As far as the individual sector queues are concerned, the disk behaves like the server with vacation and relaxation that was analysed above. Whenever the queue is empty, the server takes a vacation, equal to one rotation of the disk. If a job is served, taking a constant time τ/N, then the server is unavailable to serve the next request until a relaxation period has elapsed. The relaxation period is $(N-1)\tau/N$ long. The arrival rate at each sector queue is λ/N. Substituting these values in equation (5.14), gives the mean sector queue length as:

$$Q = \frac{\lambda\tau}{N^2} + \frac{(\lambda/N)^2\tau^2}{2(1-\lambda\tau/N)} + \frac{(\lambda/N)^2\tau}{2}$$

and the mean delay to requests as:

$$W = \frac{\tau}{N} + \frac{\lambda\tau^2}{2(N-\lambda\tau)} + \frac{\lambda\tau}{2N}$$

It can be seen that the slight overhead of organising individual sector queues pays handsome dividends in terms of the mean delay to disk requests. It will also be observed that the different service afforded to the first request of a busy period makes only a tiny difference.

5.6   Exercises

1. Use the tagged job methodology of section 5.3 to derive the mean delay for the $M/G/1$ queue with server vacations. (*Hint:* Arriving jobs find either a service in progress, or a vacation.)

2. Analyse the $M/G/1$ system in which the job starting a busy period receives exceptional service. That is, a job that on arrival finds an idle server has a different service time distribution from jobs that arrive to find the server busy.

    If $B_0^*(s)$ is the Laplace transform of these initialising service times, show that the generating function of queue length is:

    $$P(z) = \pi_0 \frac{B_0^*(\lambda(1-z))z - B^*(\lambda(1-z))}{z - B^*(\lambda(1-z))}$$

    Calculate π_0 and the mean queue length. Notice that there is no requirement that the exceptional service is longer than a normal service. It could be shorter.

3. Consider the $M/G/1$ queue in which arrivals are in batches whose size is random. If the size of batches has generating function $G(z)$, show that the generating function of queue length is given by:

$$P(z) = \frac{\pi_0 B^*(\lambda(1 - G(z)))(z - 1)}{z - B^*(\lambda(1 - G(z)))}$$

Calculate π_0 and the mean queue length.

4. Analyse the single server queue with server vacations, in which the server takes a single vacation, distributed as $H(t)$, at the end of a busy period. After the vacation, the server will start to serve customers straight away.

5.7 Further reading

The Pollaczek-Khintchine formula has been derived in many ways. Our first method can be found in Saaty's book[12], but he gives no source. The embedded Markov chain approach is based on the work of D.G. Kendall[8, 9]. D.R. Cox[2] derives the same formula by introducing a 'supplementary variable'. The variable he chooses is the remaining service time of the job in service. This transforms the non-Markovian process into a Markov process, which can be analysed. Kleinrock[10] is an easily obtainable source for this technique. The approach using Little's theorem to derive the mean queue length is difficult to attribute, but the graphical exposition of the expected remaining service time is found in Bertsekas and Gallager[1]. Yet another approach is given by Henderson[7], and Greenberg[6] has studied the $M/G/1$ queue without distribution assumptions.

Exercise 2 was first studied by Welch[13]. This model, using a special service for the job which starts a busy period, is surprisingly powerful and has been used for many different analyses. The finite capacity $M/G/1$ queue was analysed by Lavenberg[11].

Queues with server vacations have long been studied. This presentation is based on that by Gelenbe and Mitrani[5]. Fuhrmann[3, 4] has reported recent analyses of $M/G/1$ type systems.

5.8 Bibliography

[1] D.P. Bertsekas and R.G. Gallager. *Data Networks*. Prentice-Hall, Englewood Cliffs, NJ, 1987.

[2] D.R. Cox. The analysis of non-Markovian stochastic processes by inclusion of supplementary variables. *Proceedings of the Cambridge Philosophical Society*, 51(3):433–441, 1955.

[3] S.W. Fuhrmann. A note on the $M/G/1$ queue with server vacations. *Operations Research*, 32(6):1368–1373, November/December 1984.

[4] S.W. Fuhrmann and R.B. Cooper. Stochastic decompositions in the $M/G/1$ queue with generalized vacations. *Operations Research*, 33(5):1117–1129, September/October 1985.

[5] E. Gelenbe and I. Mitrani. *Analysis and Synthesis of Computer Systems*. Academic Press, London, 1980.

[6] I. Greenberg. Distribution-free analysis of $M/G/1$ and $G/M/1$ queues. *Operations Research*, 21(2):629–635, March/April 1973.

[7] W. Henderson. Alternative approaches to the analysis of the $M/G/1$ and $G/M/1$ queues. *Operations Research*, 15(1):92–101, January/February 1972.

[8] D.G. Kendall. Some problems in the theory of queues. *Journal of the Royal Statistical Society (Series B)*, 13(2):151–185, 1951.

[9] D.G. Kendall. Stochastic processes occurring in the theory of queues and their analysis by the method of the imbedded Markov chain. *Annals of Mathematical Statistics*, 24(3):338–354, 1953.

[10] L. Kleinrock. *Queueing Systems Volume 1: Theory*. Wiley, New York, NY, 1975.

[11] S.S. Lavenberg. The steady state queueing time distribution for the $M/G/1$ finite capacity queue. *Management Science*, 21(5):501–506, 1975.

[12] T.L. Saaty. *Elements of Queueing Theory*. McGraw-Hill, New York, NY, 1961.

[13] P.D. Welch. On a generalized $M/G/1$ queueing process in which the first customer of each busy period receives exceptional service. *Operations Research*, 12(5):736–752, September/October 1964.

Chapter 6

Queues with breakdowns

The models that have been studied in previous chapters ignore one major characteristic of real-world systems. All the systems analysed have had perfectly reliable servers which never breakdown. Unfortunately this desirable property is not found in the real world since all devices fail more or less frequently. They may either fail completely and provide no service, or fail partially and still provide service albeit at a reduced rate. We shall only deal with systems that are repaired when they break down.

The earlier models may provide a good approximation to the steady state of a system with breakdowns during one of its operative phases, provided that the periods of breakdown are short enough that very few jobs arrive during a breakdown. If too many jobs arrive during a broken down interval, then the server will have a large initial queue, and steady state may not be reached before the next breakdown.

In this chapter we study the simplest model, that of an $M/M/1$ queue with exponentially distributed operative and inoperative periods. The analysis is then extended to allow generally distributed service requirements for the jobs, and generally distributed repair times.

6.1 $M/M/1$ queue with breakdowns

This is the simplest example of a system which has breakdowns. The single server goes through a cycle of being available to serve jobs and then being unavailable whilst it is being repaired. It remains operative for periods which are exponentially distributed with mean length $1/\gamma$. When it breaks down, a repair process starts which is exponentially distributed with mean length $1/\eta$. Both breakdown and repair rates are independent of the number of jobs present in the system. The memoryless properties of the exponential distribution imply that the time since the last breakdown or repair does not affect the probability of a breakdown happening or the repair process completing. The steady state probability that the server is operational at any instant is given by $\eta/(\gamma + \eta)$ and that it is undergoing repair equals $\gamma/(\gamma + \eta)$.

We assume that customer arrivals form a Poisson stream with rate λ, and that the service requests are exponentially distributed in length with parameter μ, and hence have a mean service time of $1/\mu$. A service which is interrupted by a breakdown, will be resumed at the point of interruption following repair of the server. The memoryless property of the exponential distribution means that the distribution of service time remaining is identical after a repair to that which was encountered when the job started in service.[†] The system will reach a steady state if and only if the average amount of work which is brought into the system, per unit time, $\rho \overset{\triangle}{=} \lambda/\mu$, is less than the average proportion of time that the system is operational, $\eta/(\gamma + \eta)$. The state of the system can be represented by a pair of integers, (i, j), with i representing the number of operative servers (0 or 1) and j representing the number of jobs in the system. Assuming that steady state conditions will be reached, we denote the probability of state (i, j) by $\pi_{i,j}$. Consider the states in which the server is broken down ($i = 0$). The rate of leaving one of these states is $(\lambda + \eta)$; λ is the rate at which extra jobs arrive in the system (note that the server will still be broken), and η is the rate of repairing the server. Arrivals in these states occur from the state with one fewer jobs and the server broken, and when the server breaks down, in which case the number in the system remains unchanged. Converting this description to balance equations, we get:

$$(\lambda + \eta)\pi_{00} = \gamma\pi_{10} \tag{6.1}$$

$$(\lambda + \eta)\pi_{01} = \gamma\pi_{11} + \lambda\pi_{00} \tag{6.2}$$

$$(\lambda + \eta)\pi_{0j} = \gamma\pi_{1j} + \lambda\pi_{0j-1} \quad j = 2, 3, \ldots \tag{6.3}$$

When we consider the states in which the processor is available, ($i = 1$), we find another set of balance equations:

$$(\lambda + \gamma)\pi_{10} = \mu\pi_{11} + \eta\pi_{00} \tag{6.4}$$

$$(\lambda + \mu + \gamma)\pi_{11} = \lambda\pi_{10} + \mu\pi_{12} + \eta\pi_{01} \tag{6.5}$$

$$(\lambda + \mu + \gamma)\pi_{1j} = \lambda\pi_{1j-1} + \mu\pi_{1j+1} + \eta\pi_{0j} \quad j = 2, 3, \ldots \tag{6.6}$$

These equations must be satisfied by the steady state probabilities, but it is not immediately clear how to find a solution. The key is to reduce this infinite set of equations to a finite set. We define two generating functions, $G_0(z)$ and $G_1(z)$, as:

$$G_i(z) = \sum_{j=0}^{\infty} \pi_{ij}z^j, \quad \text{for } i = 0, 1$$

Taking each of equations (6.1–6.3), and multiplying them by z^j when the left hand side contains π_{0j}, we can then add them and find:

[†]The same analysis would apply if the job which was interrupted by a breakdown had to start its resumed service at the beginning, so long as the new service time was a resampling of the exponential distribution. If an interrupted job started with the same service time as before the interruption, then intuitively, service would be degraded, since a job with a long service time would be more likely to suffer an interruption than one with a short service time, and because of its long service time would be more likely to suffer a subsequent breakdown.

$$(\lambda + \eta)G_0(z) = \gamma G_1(z) + \lambda z G_0(z)$$

The manipulation of the equations (6.4–6.6) is slightly more complicated. Taking the same approach, we find that:

$$(\lambda + \mu + \gamma)G_1(z) - \mu\pi_{10} = \mu \sum_{j=1}^{\infty} \pi_{1j}z^{j-1} + \eta G_0(z) + \lambda z G_1(z)$$

and hence:

$$(\lambda + \mu + \gamma)G_1(z) = \eta G_0(z) + \mu\frac{G_1(z) - \pi_{10}}{z} + \mu\pi_{10} + \lambda z G_1(z)$$

There are three unknowns, $G_0(z), G_1(z)$, and π_{10}. Treating π_{10} as known, we can rewrite the equations as:

$$(\lambda(1 - z) + \eta)\, G_0(z) - \gamma G_1(z) = 0$$
$$-\eta z G_0(z) + \left(-\lambda z^2 + (\lambda + \mu + \gamma)z - \mu\right)G_1(z) = \mu(z - 1)\pi_{10}$$

and express both functions in terms of π_{10} and z. If we define $\mathbf{G}(z)$ as the vector $(G_0(z), G_1(z))^{\mathrm{T}}$, $\mathbf{b}$ as the vector $(0, \mu\pi_{01})^{\mathrm{T}}$, and $A(z)$ as the matrix:

$$\begin{pmatrix} \lambda(1-z) + \eta & -\gamma \\ -\eta z & -\lambda z^2 + (\lambda + \mu + \gamma)z - \mu \end{pmatrix}$$

then the equations can be expressed in matrix form as:

$$A(z)\mathbf{G}(z) = (z - 1)\mathbf{b}$$

Using Cramer's rule, we can then express each $G_i(z)$ as:

$$|A(z)|G_i(z) = (z - 1)|A_i(z)|$$

where $A_i(z)$ is the matrix $A(z)$ with the ith column replaced by $\mathbf{b}$. It is readily seen that $|A(z)|$ has a root at $z = 1$, so we can write:

$$Q(z)G_i(z) = |A_i(z)| \tag{6.7}$$

where:

$$Q(z) = (\mu - \lambda z)(\lambda(1 - z) + \eta) - \lambda\gamma z$$

Now using the normalising condition, $G_0(1) + G_1(1) = 1$, and substituting the expressions for $G_i(z)$ (equation (6.7)), evaluated at $z = 1$, we find:

$$\pi_{10} = \frac{\gamma}{\gamma + \eta}[(\mu - \lambda)\eta - \lambda\gamma]$$

From this we find expressions for $G_0(z)$, $G_1(z)$, and hence for $G(z)$, the generating function of the number in the system.

$$G(z) = G_0(z) + G_1(z)$$
$$= \frac{[(\mu - \lambda)\eta - \lambda\gamma][\gamma + \lambda(1 - z) + \eta]}{(\gamma + \eta)([\lambda(1 - z) + \eta][\mu - \lambda z] - \gamma\lambda z)}$$

The mean number in the system, independent of the state of the server, can now be easily obtained by differentiation of $G(z)$,

$$G'(1) = \frac{\lambda\left[(\gamma + \eta)^2 + \mu\gamma\right]}{(\gamma + \eta)[\eta(\mu - \lambda) - \lambda\gamma]}$$

The approach given here to solving a system with servers liable to breakdowns is a good example of the power of generating functions in queueing system analysis. It can be extended to a system of N identical processors, although a more complex analysis is needed to evaluate the $G_i(z)$ functions.

6.2 $M/G/1$ queue with breakdowns

The obvious objection to the analysis in the previous section is that the exponential distribution is unrealistic. Certainly, the repair time distribution might be anything but exponential. An obvious example occurs when the repair is effected by installing a new server, which might take a constant time. The variance of service times affects the mean queue length in an ordinary $M/G/1$ queue. What is its effect when the server can break down?

As in the previous section, we assume that arrivals are Poisson, at rate λ, and that breakdowns occur after a working interval which is exponentially distributed with parameter γ. The service time of messages is arbitrarily distributed with mean $\bar{S}$, and second moment $E[S^2]$. Similarly, repair times have an arbitrary distribution with mean $\bar{R}$, and second moment $E[R^2]$. As before, the breakdown and repair process does not affect the arrival process, so the steady state probability that the server is operational is given by $1/(1 + \gamma\bar{R})$, and the probability of being broken is $\gamma\bar{R}/(1 + \gamma\bar{R})$. These are denoted by p_1 and p_0, respectively, with the subscript representing the number of operational servers. Assuming that a steady state is reached, the proportion of time that the server is actively serving jobs is given by $\rho \triangleq \lambda\bar{S}$, assuming that interrupted jobs resume at the point of interruption, since ρ is the amount of work which enters the system per unit time. Hence the proportion of time that a job is *not* receiving service is given by $1 - \rho$. However, some of this time is unavailable because the server is broken down. The system will only be stable if there is a positive probability that the server is both idle and operational, that is:

$$\frac{1}{1+\gamma\bar{R}} > \rho$$

We denote the steady state probability that there are j jobs in the system, when the server is operational by $\pi_{1,j}$, and when it is broken by $\pi_{0,j}$. Clearly, $p_i = \sum_{j=0}^{\infty} \pi_{ij}$. The newly arriving customer sees the steady state, and so encounters an operative server with probability p_1, and a broken server with probability p_0.

Consider a job which arrives to an operational server. The server is busy with probability ρ in steady state. Conditional on the arrival finding an operational server, the probability that it will have to queue is ρ/p_1. The time until the new arrival leaves the system is made up of the remaining service time of the job in service, if any, the service times of any jobs already queueing, and its own service time, *plus* the time during which the server was under repair. Now breakdowns occur after an operational interval which is exponentially distributed. If the service of a job lasts for t, then the expected number of breakdowns that will occur is γt. Thus the expected time from the job with service requirement t starting service until it finishes service is $t(1+\gamma\bar{R})$ since each breakdown takes on average $\bar{R}$ to be repaired. Now the probability that a job which arrives at the server and finds it operational, but finds no other jobs waiting or in service, is given by $(1-\rho)/p_1$. Thus the expected waiting time, given that the arrival finds an operative server is given by:

$$W_1 = \left(\frac{(1-\rho)}{p_1}\bar{S} + \frac{\rho}{p_1}\left(\frac{E[S^2]}{2\bar{S}} + N_1\bar{S}\right)\right) \cdot \frac{1}{p_1}$$

where N_1 is the mean number in the system when the server is operative.

Now consider the waiting time encountered by a job which arrives when the server is inoperative. It will first have to wait for the residual life of the repair interval, and then for the service of all the jobs before it in the queue, and then for its own service before it can leave the system. Breakdowns do not occur while a repair is in progress, but the service of other jobs may be interrupted by further breakdowns. The service time of all the jobs in front of the arrival is not quite so simple as in the operative server case. If the first job in the queue was in service when a breakdown occurred, then it will only need a residual service time, mean $E[S^2]/(2\bar{S})$. If it arrived to find the queue empty, but the server broken down, then it will need a complete service time, mean $\bar{S}$. What are the probabilities of those two events? The server breaks down when the system is empty with probability $(1-\rho)/p_1$ and when the system is non-empty with probability ρ/p_1. Hence the probability of an arrival, during a broken down period, finding a customer who had his service interrupted is given by $p_0\rho/p_1$. The probability that a new arrival will find a queue of waiting customers and a broken down server is just $\sum_{j=1}^{\infty} \pi_{0,j} = p_0 - \pi_{0,0}$. Hence the probability that the arrival to the broken down server finds a job which had its service interrupted is given by:

$$\beta = \frac{p_0 \rho}{p_1 \cdot (p_0 - \pi_{0,0})}$$

Combining these expressions, we find that the expected waiting time of those jobs which find the server broken:

$$W_0 = \frac{E[R^2]}{2\bar{R}} + \left(N_0\bar{S} + (\beta\frac{E[S^2]}{2\bar{S}} + (1-\beta)\bar{S}) \cdot \frac{p_0 - \pi_{0,0}}{p_0} + \frac{\pi_{0,0}}{p_0}\bar{S} \right) \cdot \frac{1}{p_1}$$

This simplifies to:

$$W_0 = \frac{E[R^2]}{2\bar{R}} + \left(\frac{E[S^2]}{2\bar{S}}\frac{\rho}{p_1} + \bar{S}\frac{p_1 - \rho}{p_1} + N_0\bar{S} \right)\frac{1}{p_1}$$

It is tempting to attempt to apply Little's theorem to the system when the server is operative, and the system when the server is broken, individually. However, the theorem does *not* apply in that case, because W_0 and N_0 are not the limiting values of the means as the observed interval becomes longer and longer. The unconditional mean waiting time is given by $W = p_0 W_0 + p_1 W_1$, and the unconditional mean number in the system is given by $N = p_0 N_0 + p_1 N_1$. These unconditional mean waiting times and mean number in the system *do* satisfy Little's theorem, so after a little algebra the mean number in the system is found to be:

$$N = \frac{\rho}{p_1} + \frac{\lambda p_1 p_0 \frac{E[R^2]}{2\bar{R}} + \lambda\rho\frac{E[S^2]}{2\bar{S}\cdot p_1}}{p_1 - \rho}$$

6.3 Exercises

1. Investigate the $M/M/1$ queue with breakdowns. How does the analysis change if the breakdown can only occur while the server is operational? Does this change the conditions under which steady state will be reached?
2. Generalise the result of the previous exercise to the case in which the breakdown and repair rates are different when the system is empty.
3. Show that the $M/G/1$ system with breakdowns has mean waiting times minimised by reducing the variance of repair times.

6.4 Further reading

The analysis of queueing systems in which the server can break down has a long history. An $M/M/1$ queue with breakdowns was first studied by White and Christie[11]. The presentation here is based on the general $M/M/c$ system with breakdowns analysed by Mitrani and Avi-Itzhak[8]. Gaver[2] analysed the $M/G/1$ system with breakdowns. Our presentation follows closely the development in Avi-Itzhak and Naor[1].

Recent research on the problem of queueing systems with breakdowns has concentrated on the idea of *performability*, an attempt to combine performance and reliability in a single metric. Meyer[6, 7] coined the term and has analysed a number of types of system. Trivedi and his co-workers are also very active, using the techniques of Markov reward processes[10, 9, 5] and the *perceived* mean as a metric[4, 3].

6.5 Bibliography

[1]   B. Avi-Itzhak and P. Naor. Some queueing problems with the service station subject to breakdown. *Operations Research*, 11(3):303–320, May/June 1963.

[2]   D.P. Gaver. A waiting line with interrupted service, including breakdowns. *Journal of the Royal Statistical Society (Series B)*, 24:73–90, 1962.

[3]   R. Geist, D.E. Stevenson, and R.A. Allen. The perceived effect of breakdown and repair on the performance of multiprocessor systems. *Performance Evaluation*, 6(4):249–260, 1986.

[4]   R. Geist and K.S. Trivedi. The integration of user perception in the heterogeneous $M/M/2$ queue. In A.K. Agrawala and S.K. Tripathi, editors, *Performance '83*, pages 203–216. North Holland, Amsterdam, 1983. Proceedings of Performance '83, 9th International Symposium of Computer Performance Modeling, Measurement, and Evaluation, College Park, University of Maryland, 25–27 May 1983.

[5]   V.G. Kulkarni, V.F. Nicola, and K.S. Trivedi. On modelling the performance and reliability of multimode computer systems. *The Journal of Systems and Software*, 6(1/2):175–182, May 1986.

[6]   J.F. Meyer. On evaluating the performability of degradable computer systems. *IEEE Transactions on Computers*, C-29(8):720–731, January 1980.

[7]   J.F. Meyer. Closed-form solution of performability. *IEEE Transactions on Computers*, C-31(7):648–657, July 1982.

[8]   I. Mitrani and B. Avi-Itzhak. A many-server queue with service interruptions. *Operations Research*, 16(3):628–638, May/June 1968.

[9]   V.F. Nicola, V.G. Kulkarni, and K.S. Trivedi. Queueing analysis of fault tolerant computer systems. *IEEE Transactions on Software Engineering*, SE-13(3):363–375, March 1987.

[10] R.M. Smith, K.S. Trivedi, and A.V. Ramesh. Performability analysis: Measures, an algorithm, and a case study. *IEEE Transactions on Computers*, C-37(4):406–417, April 1988.

[11] H.C. White and L.S. Christie. Queuing with preemptive priorities or with breakdown. *Operations Research*, 6(1):79–95, January/February 1958.

Chapter 7

Priority queues

In this chapter we shall examine the effect of giving some jobs priority over others. For example, if the processing requirements of each job were known in advance, jobs could be served in order of increasing processing time. In other cases some external priority may be imposed, unrelated to the processing time. A professor's job might be given priority over student work, regardless of their respective processing times. The two types of priority are sometimes distinguished by calling the first a scheduling discipline. We shall not make this distinction. A more descriptive distinction is to call the first system *internal priorities* and the second *external priorities*.

Both types of priority system fall into two broad classes, *preemptive* and *non-preemptive*. In a non-preemptive system, the priority rule is only used when a new job is to be allocated to the processor after an idle period or when the previous job has finished processing. The job which is being processed is never interrupted. In a preemptive system, on the other hand, a job which has higher priority than the job which is currently executing will take over the processor, displacing the previous job. In a preemptive system, jobs of the highest priority are unaffected by the presence of jobs of other priority classes; in a non-preemptive system, they will still be affected by the presence of low priority jobs because they must wait for the residual service time of the job being served before they can be processed. Conventionally, different priority classes of jobs are given an index, with the lowest index being assigned to the jobs which have highest priority.

7.1 Non-preemptive priority queue

We start by analysing the $M/G/1$ queue with a non-preemptive priority discipline. We assume that there are R classes of job, and that jobs of class i have priority over class j if $i < j$. Jobs of class i arrive in a Poisson stream at rate λ_i, and have service times which are generally distributed with mean $1/\mu_i$ and second moment M_{2i}. The average rate at which work from class i jobs is introduced into the system is $\rho_i = \lambda_i/\mu_i$, and the total rate at which work arrives is $\sum_{i=1}^{R} \rho_i$. If this total rate exceeds 1, then at least some priority classes will not reach steady-state. For the moment we assume that the total arrival rate is such that steady state will be reached. In that case, ρ_i is the probability that the processor is serving a job of class i. This can be verified by using Little's theorem. Define the system consisting of only the server, and considering only class i jobs, they arrive at rate λ_i, and have mean residence time $1/\mu_i$. Hence the mean number of class i jobs being served is ρ_i. Since arrivals form a Poisson process, they observe the same system as a random observer. The mean residual service time for a class i job is found using equation (5.11) and equals $\frac{1}{2}\mu_i M_{2i}$, so the overall expected residual service time of a job in service is:

$$W_0 = \frac{1}{2}\sum_{i=1}^{R}\lambda_i M_{2i} \tag{7.1}$$

The mean residence time of jobs of class i is found by considering the time that a tagged job will spend in the system.

Consider first a job of class 1, the highest priority. Its residence time is made up of the residual service time of the job in service when it arrives, if any, the service times of the jobs of class 1 that it finds in the queue, and its own service time. Taking expectations we find:

$$W_1 = W_0 + Q_1/\mu_1 + 1/\mu_1$$

where W_1 is the mean residence time of a class 1 job, and Q_1 the mean number of class 1 jobs in the queue, not including any job being processed. If N_1 is the mean number of class 1 jobs in the system, including the job in service, if any, then it follows that:

$$Q_1 = N_1 - \rho_1$$

since with probability ρ_1 a job of class 1 is being processed. Using Little's theorem,

$$W_1 = W_0 + \frac{\lambda_1 W_1 - \rho_1}{\mu_1} + \frac{1}{\mu_1}$$

or:

$$W_1 = \frac{1}{\mu_1} + \frac{W_0}{1 - \rho_1}$$

mean waiting time Now consider class 2 jobs. Any delay incurred by a class 2 job before it enters service potentially allows more class 1 jobs to arrive and be served first. Quantifying this extra delay, we note that, on average, in time T, $\lambda_1 T$ new jobs of class 1 will enter the system, each with a mean service requirement of $1/\mu_1$ units. $\rho_1 T$ is the expected amount of class 1 processing that enters the system in time T. Thus any delay of T to a class 2 job will be expanded by a further $\rho_1 T$ caused by the class 1 jobs that arrive in the interim. During this extra $\rho_1 T$ delay, a further $\rho_1^2 T$ units of class 1 work enter the system, and so on. In general, the total delay will be:

$$T(1 + \rho_1 + \rho_1^2 + \rho_1^3 \cdots) = T/(1 - \rho_1)$$

Now the initial delay encountered by a class 2 job is composed of the residual service time of the job in service, the service times of all the class 1 and class 2 jobs in the queue at the time of the job's arrival. To this initial delay must be applied the expansion factor of $1/(1 - \rho_1)$ which allows for the extra class 1 jobs that enter, and then the job's own service time added to give:

$$W_2 = \left[W_0 + \frac{N_1 - \rho_1}{\mu_1} + \frac{N_2 - \rho_2}{\mu_2} \right] / (1 - \rho_1) + \frac{1}{\mu_2}$$

Since $N_1 = \lambda_1 W_1$ and $N_2 = \lambda_2 W_2$, and we already know W_1, we can solve this equation and find:

$$W_2 = \frac{1}{\mu_2} + \frac{W_0}{(1 - \rho_1)(1 - \rho_1 - \rho_2)}$$

This same approach could be followed for classes 3, 4, ... A class 3 job will be delayed by class 1 and class 2 jobs, which bring new work into the system at rate $\rho_1 + \rho_2$ per unit time. The 'raw' delay which a class 3 job encounters is made up of the residual service time of the job in service, and the service times of all class 1, 2, and 3 jobs in the system when it arrives. This 'raw' delay is expanded by the factor $1/(1 - \rho_1 - \rho_2)$. In general, observe that as far as jobs in class j are concerned, there is only one class of jobs with higher priority, which has traffic intensity $\rho_H = \sum_{i=1}^{j-1} \rho_i$, and the presence of lower priority classes only influences the residual service time W_0. Thus from the point of view of class j jobs, there are only 2 classes, H and j, and the formula for the mean waiting time is:

$$
\begin{aligned}
W_j &= \frac{1}{\mu_j} + \frac{W_0}{(1 - \rho_H)(1 - \rho_H - \rho_j)} \\
&= \frac{1}{\mu_j} + \frac{W_0}{\left(1 - \sum_{i=1}^{j-1} \rho_i \right) \left(1 - \sum_{i=1}^{j} \rho_i \right)}
\end{aligned}
$$

7.2   Preemptive priority queues

In the non-preemptive case, once a job had started service, it continued to the end, without interruption. If we allow higher priority jobs to interrupt service and take over the processor, we have a preemptive priority discipline. It is important to distinguish three different preemptive disciplines depending on the fate of the interrupted job: it can be resumed from the point of interruption, without the loss of any work already done on it; it can be restarted from the beginning with the same length as before; or it can be started from the beginning with a new length drawn from the appropriate distribution. These three choices for what to do with the interrupted job are known as *preemptive resume, preemptive restart without resampling*, and *preemptive restart with resampling*, respectively. Intuitively, preemptive resume will have the lowest mean response time. The effect of resampling is less clear cut, but it seems reasonable that it will lead to lower mean response times than those achieved by not resampling. The reasoning is the following: a long job is more likely to be interrupted than a short one, and if the same job is restarted it is still likely to be interrupted, whereas if it is resampled, there is a good chance that the new length will be less, giving a better chance of completion.

We shall not analyse preemptive restart disciplines here, only preemptive resume. Clearly jobs of class 1 are unaffected by the presence of other job classes in the system. Thus,

$$W_1 = W_{01} + \frac{Q_1}{\mu_1} + \frac{1}{\mu_1}$$

where W_{01} is the mean residual life of jobs of class 1, and equals $\frac{1}{2}\lambda_1 M_{21}$. This is exactly the same formula as in the non-preemptive case, except that the residual life only includes jobs of class 1. Hence,

$$W_1 = \frac{W_{01}}{(1 - \rho_1)} + \frac{1}{\mu_1}$$

Jobs of class 2 have to wait for the residual service time of the job in service, if it is of class 1 or 2, then for the service times of any jobs of classes 1 or 2 which are already waiting, as well as any class 1 jobs which arrive while it is waiting, and finally receive its own service, albeit interrupted by class 1 arrivals. The service of class 1 and 2 jobs which are already waiting and the class 1 jobs which arrive during that service is exactly the same as it was in the non-preemptive case. However the term representing the job's own service is not $1/\mu_2$, because this must include time used by class 1 jobs which arrive during the service of this tagged job and preempt it. By the same argument used above, this introduces an expansion factor of $1/(1 - \rho_1)$, so the waiting time satisfies:

$$W_2 = \frac{1}{(1 - \rho_1)}\left(\frac{1}{\mu_2} + \frac{W_{01} + W_{02}}{(1 - \rho_1 - \rho_2)}\right)$$

where W_{02} is the residual service time of a class 2 job. Again we observe that this generalises in a straightforward manner to the response time for class i jobs,

$$W_i = \frac{1}{(1 - \sum_{j=1}^{i-1}\rho_j)}\left(\frac{1}{\mu_i} + \frac{\sum_{j=1}^{i}W_{0j}}{(1 - \sum_{j=1}^{i}\rho_j)}\right)$$

7.3 Kleinrock's conservation law

The previous sections of this chapter have shown the effect of giving priority to different jobs. We now ask ourselves what are the boundaries of the performance which are achievable using priority schemes on a single server. To minimise the waiting time of a particular class of jobs, we give that class top priority. However this will increase the waiting time of other classes of job and we would like to know what the overall effect will be.

The unfinished work in a queue which has no priority scheme, or only a non-preemptive priority scheme, is the total service required by jobs which have not started their service, plus the remaining service time of the job in service. For preemptive priority schemes it also includes the remaining service times of those jobs which have received some service but which have been preempted. Denote the unfinished work at time t by $U(t)$. The value of $U(t)$ for a single server queue is not influenced by the order of service, so long as the service discipline is conservative. It will not be influenced by a preemptive-resume scheme either. The behaviour of the function is simple to describe; apart from arrival instants it decreases at rate 1 unit per time unit until it reaches 0, when the server is idle; it remains at 0 until the next arrival. At arrival instants, $U(t)$ increases by the service of the customer which has just arrived. At any time, $U(t)$ is the residual service time of the job in service plus the sum of the service times of all the jobs waiting in the queue.

Consider a non-preemptive priority, $M/G/1$, queue. Assume that we have R classes of job, and that there are $n_r(t)$ jobs of class r waiting at time t, with service times s_{ir}, $i = 1, \ldots, n_r(t)$. From the description above, writing x_0 for the residual service time of the job in service, if any, it follows that:

$$U(t) = x_0 + \sum_{r=1}^{R} \sum_{i=1}^{n_r(t)} s_{ir}$$

Taking expectations of both sides, and using the law of total probability,

$$E[U(t)] = W_0 + \sum_{r=1}^{R} \Pr[n_r(t) = n_r] \sum_{i=1}^{n_r} E[s_{ir}]$$

where W_0 is the expected remaining service time of the job in service, which is given by equation (7.1). But $\sum_{i=1}^{n_r} E[s_{ir}] = n_r \bar{s}_r$, since the expected service time does not depend on the position in the queue. Taking limiting values as $t \to \infty$, and using N_r for the mean number of class r jobs *queueing*, and V_r for the mean queueing time of class r jobs,

$$\bar{U} = W_0 + \sum_{r=1}^{R} N_r \bar{s}_r$$

$$= W_0 + \sum_{r=1}^{R} \lambda_r V_r \bar{s}_r$$

$$= W_0 + \sum_{r=1}^{R} \rho_r V_r$$

Since we know that $\bar{U}$ is independent of the priority discipline, we can calculate it using a non-priority $M/G/1$ model. The value of the mean queueing time, V, was calculated in equation (5.12).

$$\bar{U} = W_0 + \rho V$$
$$= W_0 + \rho \frac{W_0}{1 - \rho}$$
$$= \frac{W_0}{1 - \rho}$$

Hence,

$$\sum_{r=1}^{R} \rho_r V_r = \frac{\rho W_0}{1 - \rho}$$

This conservation law enables us to calculate the bounds of the achievable performances with a non-preemptive priority scheme on a single server. Since the weighted sum of the waiting times remains constant, any decrease in one class's waiting time must be paid for by an increase in some other class's waiting time.

7.4   Service time dependent priorities

The analysis of priorities given above does not impose any restriction on how incoming jobs are assigned to priority classes. If the execution times are known in advance, as would be the case in most communication systems, since message lengths are known, then the jobs can be grouped into classes according to length of service required. This would enable us to give priority to short messages, at the expense of longer waiting times for the other messages. In many packet switching networks, the packets have a bi-modal length distribution, that is, there are many very short packets, a product of the interactive terminal traffic being carried by the network, and also many long packets, of the maximum length allowed by the network protocols, which are the product of file transfer and similar less urgent traffic. These two packet lengths, the minimum and the maximum make up a high proportion of the traffic.

For the sake of a simple example, assume that no other packet lengths exist, and that the short packets make up proportion p of the traffic, and have transmission time t_S, and that the long packets, which form the remaining $1 - p$ of the traffic have transmission time t_L. Hence the mean service time for a packet is $pt_S + (1 - p)t_L$, and the second moment is $pt_S^2 + (1 - p)t_L^2$. The packets arrive at rate λ per second. The traffic intensity, ρ, is $\lambda(pt_S + (1 - p)t_L)$. If the packets are served in strictly first-come-first-served order, then the Pollazcek-Khintchine formula applies, and the mean queueing time for a packet is given by:

$$V = \frac{\lambda(pt_S^2 + (1 - p)t_L^2)}{2(1 - \rho)}$$

If the short messages are given higher priority , then their queueing time is given by:

$$V_S = \frac{\lambda(pt_S^2 + (1 - p)t_L^2)}{2(1 - \lambda pt_S)}$$

The long packets will have mean queueing time:

$$V_L = \frac{\lambda(pt_S^2 + (1 - p)t_L^2)}{2(1 - \lambda pt_S)(1 - \rho)}$$

As might be expected, the three queueing times satisfy the following order relationship:

$$V_S < V < V_L$$

If the longer packets are given priority, then:

$$V_L = \frac{\lambda(pt_S^2 + (1 - p)t_L^2)}{2(1 - \lambda(1 - p)t_L)}$$
$$V_S = \frac{\lambda(pt_S^2 + (1 - p)t_L^2)}{2(1 - \lambda(1 - p)t_L)(1 - \rho)}$$

and the relationship between V_S, V, and V_L is reversed in sense.

7.5   Exercises

1. Consider a two class preemptive priority system. High priority jobs have exponentially distributed service times with mean $1/a$. Low priority jobs have exponentially distributed service times with mean $1/b$. Both classes arrive at equal rate λ in Poisson streams. Compare the performance of such a system with a two class system in which arrivals occur at rate 2λ and are assigned to the high priority class if their service request is shorter than a threshold of σ seconds, and otherwise become low priority jobs. Assume that the overall distribution of service times is exponential with mean $2/(a + b)$.

7.6 Further reading

Priority queues have a long history. Cobham[3] gave the first analysis of priority queues, and derived the mean waiting time in non-preemptive priority queues. Other authors also early in the field were White and Christie[12], Phipps[10], and Avi-Itzhak and Naor[2]. The conservation law was first proved by Kleinrock[8], and an alternative proof has been given by Schrage[11].

The book by Jaiswal [7] is a comprehensive account of priority queueing analysis.

J.R. Jackson analysed queues in which the priority structure is not static, but depends on the waiting time of the jobs in the system, older jobs receiving higher priority[4, 5, 6]. Kleinrock has also analysed similar time-dependent priority systems[9].

Using delay cycle techniques, developed in chapter 8, it is possible to analyse systems in which the priority classes alternate [1].

7.7 Bibliography

[1] B. Avi-Itzhak, W.L. Maxwell, and L.W. Miller. Queueing with alternating priorities. *Operations Research*, 13(2):306–318, March/April 1965.

[2] B. Avi-Itzhak and P. Naor. On a problem of preemptive priority queueing. *Operations Research*, 9(5):664–672, September/October 1961.

[3] A. Cobham. Priority assignment in waiting line problems. *Operations Research*, 2(1):70–76, January/February 1954.

[4] J.R. Jackson. Some problems with queueing with dynamic priorities. *Naval Research Logistics Quarterly*, 7(3):235–249, 1960.

[5] J.R. Jackson. Queues with dynamic priority discipline. *Management Science*, 8(1):18–34, 1961.

[6] J.R. Jackson. Waiting-time distributions for queues with dynamic priorities. *Naval Research Logistics Quarterly*, 9(1):31–36, 1962.

[7] N.K. Jaiswal. *Priority Queues*. Academic Press, London, 1968.

[8] L. Kleinrock. A conservation law for a wide class of queueing disciplines. *Naval Research Logistics Quarterly*, 12(2):181–192, 1965.

[9] L. Kleinrock and R.P. Finkelstein. Time dependent priority queues. *Operations Research*, 15(1):104–116, January/February 1967.

[10] T.E. Phipps. Machine repair as a priority waiting line problem. *Operations Research*, 4(1):76–85, January/February 1956.

[11] L.E. Schrage. An alternative proof of a conservation law for the queue $G/G/1$. *Operations Research*, 18(1):185–187, January/February 1970.

[12]   H.C. White and L.S. Christie. Queuing with preemptive priorities or with breakdown. *Operations Research*, 6(1):79–95, January/February 1958.

Chapter 8

Waiting time distributions of queues

In this chapter we introduce a powerful method for calculating the distributions of several of the important time intervals in a queue. The busy period is the time that a server is operative. For a single server we know that the proportion of time the server is busy is equal to the traffic intensity. This technique allows us to calculate the actual distribution of the length of a busy period. We also calculate the waiting time distribution using similar techniques. Throughout this chapter we shall be considering an $M/G/1$ queue. The arrival stream is Poisson, with rate λ. The single server provides a service to jobs which has distribution function $B(t)$, and Laplace transform $B^*(s)$. The mean service time requirement is $\bar{S}$, and the traffic intensity, $\rho = \lambda \bar{S}$.

8.1 Waiting time distribution

We have seen that the mean delay to customers in a queue can be found by evaluating the mean number in the queue and using Little's theorem. However this approach tells us nothing about the distribution of the delay, only its mean.

Any discipline that chooses customers for service independently of their required service time will have the same distribution of the number of customers in the queue. This can be proved by observing that the argument that led to $P(z)$, the generating function of the number in the queue, in chapter 5, does not specify the discipline to be used to choose the next job for service. What will be the distribution of the waiting time under different disciplines? We can easily derive it for FCFS, but other disciplines are more complex.

8.2 FCFS waiting time

In a FCFS system, the jobs left in the queue are just those jobs which have arrived during the waiting time of the job which has just been served. Let us assume we know the distribution function of the waiting times, $W(t)$, and its Laplace transform, $W^*(s)$. Consider a departing job, which has been in the system for a time t. It leaves behind i jobs, all of which arrived during its residence in the system, and since arrivals are Poisson, the distribution of i satisfies $\mathrm{Pr}(\text{No of jobs left} = i) = \mathrm{e}^{-\lambda t}(\lambda t)^i/i!$. Hence the generating function of the number of jobs left by a departing job whose system residence time was t is given by:

$$P(z,t) = \mathrm{e}^{-\lambda t(1-z)}$$

In order to find the generating function of the number left in the system by a random departing job, independent of its residence time, we must integrate with respect to the system residence time distribution function, $W(t)$.

$$\begin{aligned} P(z) &= \int_0^\infty P(z,t)\,\mathrm{d}W(t) \\ &= \int_0^\infty \mathrm{e}^{-\lambda(1-z)t}\,\mathrm{d}W(t) \end{aligned}$$

which is the definition of the Laplace transform of $W(t)$. Hence $P(z)$, the generating function of the number left in the queue, is given by $W^*(\lambda(1-z))$. $P(z)$ is already known from (5.10):

$$P(z) = \frac{(1-\rho)B^*(\lambda(1-z))(1-z)}{B^*(\lambda(1-z)) - z}$$

Hence, we can find $W^*(s)$ by making the change of variable $s = \lambda(1-z)$:

$$W^*(s) = \frac{(1-\rho)sB^*(s)}{s - \lambda + \lambda B^*(s)}$$

8.3   Busy period analysis

The proportion of time that the server is busy is known from the corollary to Little's theorem. The distribution of the lengths of times that the server is busy is not known however. Is the server busy in a regular cycle, with low variability, or is there a probability of extremely long busy periods? Although the idle periods are known to be exponentially distributed with mean length $1/\lambda$, the distribution of the lengths of the busy periods is unknown. Let the busy period have probability distribution function $H(t)$, with mean length $\bar{H}$, and Laplace transform $H^*(s)$.

If $\bar{H}$ is the mean length of a busy period, then the mean length of time between the start of successive busy periods is $1/\lambda + \bar{H}$. The processor is idle during this cycle for a period of length $1/\lambda$ so the probability of finding the processor being idle is: ·

$$\frac{1/\lambda}{1/\lambda + \bar{H}}$$

but the idle probability is well known from Little's theorem corollary to be $1 - \rho$. So we find that:

$$\bar{H} = \frac{\bar{S}}{1 - \rho} \tag{8.1}$$

To find the distribution of busy period lengths, $H(t)$, first observe that the order of executing jobs will have no effect on the length of the busy period, so any convenient discipline can be chosen for the purposes of analysis. Assume that a job arrives when the processor is idle, and starts a busy period. If no further jobs arrive during its processing, then the busy period will consist of only a single service time. However, if new jobs arrive, then their service times, and those of any jobs which arrive during their processing will be part of the busy period.

Let the processing time of the initial job, J_0, be t, and assume that n jobs arrive during that period. Let these jobs be $J_1, \ldots, J_n$. Consider the following discipline: after the completion of J_0's execution take J_1 and execute it and all the jobs which arrive during its processing, only starting to execute J_2 when there are no jobs left except $J_2, \ldots, J_n$. Continue in a similar fashion, choosing one of the original jobs only when there are no later arrivals to be considered. This highly artificial discipline enables us to evaluate the busy period distribution.

Now the time to execute J_1 and all the jobs which arrive during its processing is distributed in exactly the same way as the time to execute J_2 and all those jobs which arrive in its service, and, this common distribution is just that of the busy period that we are seeking.

Assume that the service time of J_0 is t, then denoting the length of the busy period initiated by J_i by D_i:

$$E(\mathrm{e}^{-sD_0}|t, n) = E(\mathrm{e}^{-s(t+\sum_{i=1}^{n} D_i)}) = E(\mathrm{e}^{-st} \prod_{i=1}^{n} \mathrm{e}^{-sD_i})$$

$$= \mathrm{e}^{-st} H^*(s)^n$$

Removing the conditioning on n, we find:

$$E(\mathrm{e}^{-sD_0}|t) = \sum_{n=0}^{\infty} \frac{(\lambda t)^n}{n!} \mathrm{e}^{-\lambda t} \mathrm{e}^{-st} H^*(s)^n$$

$$= \mathrm{e}^{-st} \mathrm{e}^{-\lambda t} \mathrm{e}^{\lambda t H^*(s)}$$

and removing the conditioning on t gives:

$$E(\mathrm{e}^{-sD}) = H^*(s) = B^*(s + \lambda - \lambda H^*(s)) \tag{8.2}$$

This is a functional equation which usually cannot be solved to give a formula for $H^*(s)$. However, we can usually derive formulas for the moments of $H(t)$ by differentiation.

$$H^{*\prime}(0) = B^{*\prime}(\lambda - \lambda H^*(0))(1 - \lambda H^{*\prime}(0))$$

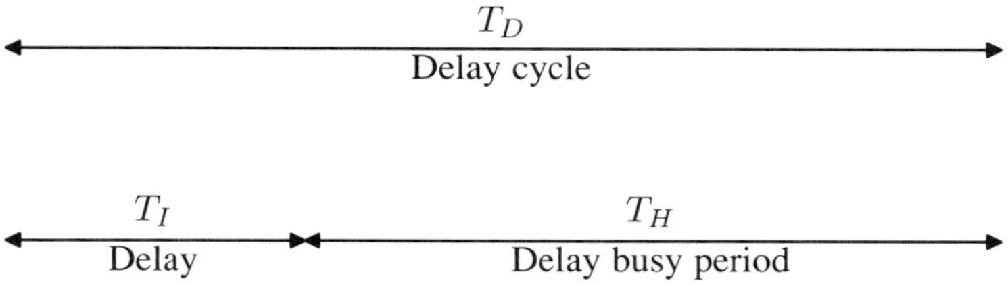

Figure 8.1 A delay cycle

Now $H^*(0) = 1, \ H^{*\prime}(0) = -\bar{H}$ and $B^{*\prime}(0) = -\bar{S}$, hence:

$$\bar{H} = \frac{\bar{S}}{1-\rho}$$

which agrees with the earlier formula for the mean (8.1). As many moments as needed can be evaluated using (8.2), at least in principle. For example, the second moment of the busy period is given by $H^{*\prime\prime}(0)$,

$$E[H^2] = \frac{E[S^2]}{(1-\rho)^3}$$

This approach, of assuming that we already know the unknown function, and deriving an equation for its Laplace transform is often employed. The key observation is the fact that the distribution of the sub-busy periods initiated by $J_1, J_2, \ldots, J_n$ is identical to the distribution of the busy period. This busy period was initiated by a normal job's service time.

In many cases, the job which starts a busy period receives special service. For example, in a circuit switched network, the first message to a particular destination causes a circuit to be established which is maintained until there are no further messages for that destination. The distribution of the length of time that the circuit is in use can be analysed in the same manner as the busy period above.

The initial job starts a set-up phase, whose distribution is known, and is then served, along with any jobs that have arrived subsequently, until the queue is empty. The time from the initiation of the set-up phase until the queue is empty again is called a *delay cycle*. The initial set-up phase is called the *delay*, and the time actually used processing jobs, the *delay busy period*. Assume that jobs arrive in a Poisson stream at rate λ, and that (non-delay) busy periods have distribution $H(t)$. The initial set-up lasts for T_I time units, with distribution $H_I(t)$. We wish to find the distribution of T_H, the length of the delay busy period, and T_D, the length of the delay cycle.

$$T_D = T_I + T_H$$

Now suppose that i jobs arrive during the initial set up. The delay busy period will be made up of i ordinary busy periods.

$$E[e^{-sT_H}|T_I = t \text{ and } N_t = i] = E[e^{-s\sum_{j=1}^{i} h_j}] = [H^*(s)]^i$$

where N_t is the number of jobs arriving in t time units, and h_j is the length of the busy period caused by the processing of the jth job. Removing the conditioning on i:

$$E[e^{-sT_H}|T_I = t] = \sum_{i=0}^{\infty} H^*(s)^i \frac{(\lambda t)^i}{i!} e^{-\lambda t} = e^{-\lambda t(1-H^*(s))}$$

and removing the conditioning on t:

$$H_H^*(s) = E[e^{-sT_H}] = \int_0^{\infty} e^{-\lambda t(1-H^*(s))} \, \mathrm{d}H_I(t) = H_I^*(\lambda(1 - H^*(s))) \quad (8.3)$$

Similarly, we can easily show that:

$$H_D^*(s) = H_I^*(s + \lambda(1 - H^*(s))) \quad (8.4)$$

This style of analysis is often called *delay cycle analysis*. Notice that although it will probably be impossible to explicitly invert the transform, we can obtain, in principle, all the moments of the distribution. The mean of the delay busy period is:

$$E[T_H] = -H_H^{*\prime}(0) = \lambda E[T_I]\bar{H} = \frac{\rho E[T_I]}{1 - \rho}$$

Similarly, the mean length of the delay cycle is given by:

$$E[T_D] = -H_D^{*\prime}(0) = E[T_I](1 + \lambda\bar{H}) = \frac{E[T_I]}{1 - \rho}$$

8.4   Waiting time distributions

We have already observed that the waiting time distribution in a FCFS system can be discovered by an elementary observation about the generating function of the number in the queue, and the way in which they arrived.

To show the power of delay cycle analysis, we now derive the waiting time distribution directly. A job arrives to find the processor idle with probability $1 - \rho$. Its time in the system, ω, will be the same as its processing time, so the Laplace transform of the waiting time distribution, $W^*(s)$, is given by:

$$W^*(s) = (1 - \rho)B^*(s) + \rho \cdot E[e^{-s\omega}|\text{busy}]$$

We need to find the time in the system of a job which arrives and has to wait. Consider a busy period of length T, and divide it into intervals T_i, as follows: let T_0 be the processing time of the job that started the busy period, and assume that N_0 jobs arrive during that interval, let T_1 be the sum of the processing times of those N_0 jobs, and assume that N_1 jobs arrive during T_1; in general, T_j is the time required to process the N_{j-1} arrivals which occurred in the previous interval, T_{j-1}. We denote the distribution of the length of interval T_j by $H_j(t)$. Assuming that the system is non-saturated, the busy period will be finite, and $T_j = 0$ for all large enough j. It follows that $\lim_{j\to\infty} H_j^*(s) = 1$.

What is the relationship between T_j and T_{j-1}? From the convolution properties of the Laplace transform,

$$E[e^{-sT_j} | T_{j-1} = t \text{ and } N_{j-1} = n] = (B^*(s))^n$$

and deconditioning on T_{j-1} and N_{j-1} gives:

$$\begin{aligned} H_j^*(s) = E[e^{-sT_j}] &= \int_0^\infty \sum_{n=0}^\infty \frac{(B(s)\lambda t)^n}{n!} e^{-\lambda t}\, dH_{j-1}(t) \\ &= H_{j-1}^*(\lambda(1 - B^*(s))) \end{aligned}$$

Consider an arrival which finds the processor busy. With probability π_j, it will arrive during the interval T_j. Assuming the queue is stable, then the queue length process is a regenerative process, with regeneration points at the start of every busy period. By the fundamental theorem, $\pi_j = E[T_j]/E[T]$. How long will it remain in the system? It must wait until the end of the current interval, a time of length R, and then for all the N jobs which arrived before it in T_j. Its system time, ω, satisfies:

$$E[e^{-s\omega} | T_j = t, N = n, R = r] = e^{-sr}(B^*(s))^{n+1}$$

Arrivals are Poisson, and the n arrivals occurred in an interval of length $t - r$, so:

$$E[e^{-s\omega} | T_j = t, R = r] = e^{-sr} B^*(s) e^{-\lambda(t-r)} \sum_{n=0}^\infty \frac{(\lambda(t-r)B^*(s))^n}{n!}$$

r is the time from a random arrival until the end of the jth interval. It is a random modification on the interval T_j; the density function of r given that $T_j = t$ is $1/t$, and the density function of t, the *observed* interval length, is $tH_j(t)/E[T_j]$.

Hence, removing the conditioning on r and t,

$$\begin{aligned} W_j^*(s) &= E[e^{-s\omega} | \text{arrival in period } j] \\ &= B^*(s) \int_0^\infty \int_0^t (e^{-t\lambda(1-B^*(s))} e^{-r(\lambda(B^*(s)-1)+s)})\, dr\, \frac{dH_j(t)}{E[T_j]} \end{aligned}$$

$$= \frac{B^*(s)}{E[T_j](\lambda(B^*(s) - 1) + s)} \int_0^\infty (e^{-t\lambda(1-B^*(s))} - e^{-st})\, dH_j(t)$$

$$= \frac{B^*(s)(H^*_{j+1}(s) - H^*_j(s))}{E[T_j](\lambda(B^*(s) - 1) + s)}$$

We now have to remove the conditioning on arrival during period j.

$$E[e^{-s\omega}|\text{busy}] = \sum_{j=0}^\infty \pi_j W^*_j(s)$$

$$= \sum_{j=0}^\infty \frac{E[T_j]}{E[T]} \frac{B^*(s)(H^*_{j+1}(s) - H^*_j(s))}{E[T_j](\lambda(B^*(s) - 1) + s)}$$

$$= \frac{B^*(s)}{E[T](\lambda(B^*(s) - 1) + s)} \sum_{j=0}^\infty (H^*_{j+1}(s) - H^*_j(s))$$

The summation term collapses by cancellation, and since $\lim_{j\to\infty} H^*_j(s) = 1$,

$$E[e^{-s\omega}|\text{busy}] = \frac{B^*(s)(1 - H^*_0(s))}{E[T](\lambda(B^*(s) - 1) + s)}$$

Now $H^*_0(s) = B^*(s)$, and $E[T]$ is the expected length of a busy period, hence:

$$W^*(s) = (1 - \rho)B^*(s) + \rho \cdot E[e^{-s\omega}|\text{busy}]$$

$$= \frac{(1 - \rho)sB^*(s)}{s - \lambda(1 - B^*(s))}$$

In the case of a delay cycle, the length of the initiating period, T_0, is the initial delay. No job arrives to find an idle server, because any job which arrives to an empty system initiates the delay period. The waiting time Laplace transform is given by:

$$W^*(s) = \frac{B^*(s)(1 - H^*_I(s))}{E[T_D](\lambda(B^*(s) - 1) + s)}$$

where $H^*_I(s)$ is the Laplace transform of the initial delay, and $E[T_D]$ is the expected length of the whole delay cycle. Equivalently, it can be expressed in terms of the length of the initial delay, to give:

$$W^*(s) = \frac{(1 - \rho)B^*(s)(1 - H^*_I(s))}{E[T_0](\lambda(B^*(s) - 1) + s)} \qquad (8.5)$$

The queueing time is simply found by dividing by $B^*(s)$, since the Laplace transform of the sum of two random variables is the product of their Laplace transforms.

8.5   Waiting time for LCFS discipline

We have observed that the distribution of the number of customers in an $M/G/1$ system is independent of the service discipline (provided that jobs are chosen without regard to their service time). This is not so for their waiting time distribution. Let us now consider the last-come-first-served (LCFS) service discipline. In this discipline, a job only starts its service when all the jobs that arrived after it have been served. The queue length distribution is the same, and hence, because of Little's result, the mean waiting time will be identical, but higher moments of the waiting time will be different.

Consider a tagged job: if it arrives to find the server idle it starts service immediately and has the same waiting time as in an FCFS discipline. If it finds the server busy it waits until the end of the service of the job currently using the server, and then for the service of all jobs which arrived after the tagged job before it can start service. The time until the current job finishes service is a random modification of the service time. Its distribution was found in section 5.4. Suppose that i jobs arrive during that interval. Each of those jobs will give rise to a busy period with the distribution function $H(t)$, that was found in section 8.3 above. Thus the queueing time of our tagged job when it finds the server busy is a delay cycle in which the initial delay is a random modification of a service time. Now the Laplace transform of the random modification of service time by (5.13) is:

$$R^*(s) = \frac{1 - B^*(s)}{s\bar{S}}$$

and substituting $R^*(s)$ for $H_I^*(s)$ in equation (8.4) gives the transform of the queueing time distribution, conditioned on the server being busy:

$$\begin{aligned}
V^*(s) &= R^*(s + \lambda(1 - H^*(s))) \\
&= \frac{1 - B^*(s + \lambda(1 - H^*(s)))}{(s + \lambda(1 - H^*(s))) \cdot \bar{S}} \\
&= \frac{1 - H^*(s)}{(s + \lambda(1 - H^*(s))) \cdot \bar{S}}
\end{aligned}$$

If the system is empty when the job arrives, the waiting time is just the service time. This occurs with probability $1 - \rho$. If queueing is necessary, then the waiting time is the sum of the service time and the queueing time. Combining these equations:

$$\begin{aligned}
W^*(s) &= (1 - \rho)B^*(s) + \rho B^*(s)V^*(s) \\
&= B^*(s)\left((1 - \rho) + \frac{\lambda(1 - H^*(s))}{s + \lambda(1 - H^*(s))}\right)
\end{aligned}$$

As we expect, the mean queueing time is identical to the FCFS case, but the second moment in the LCFS case is:

$$E[V^2] = \frac{\lambda E[S^3]}{3(1-\rho)^2} + \frac{(\lambda E[S^2])^2}{2(1-\rho)^3}$$

whereas in the FCFS case it is:

$$E[V^2] = \frac{\lambda E[S^3]}{3(1-\rho)} + \frac{(\lambda E[S^2])^2}{2(1-\rho)^2}$$

The variance of the queueing time is less under the FCFS discipline by a factor of $(1-\rho)$. If the system is heavily loaded, ρ will be close to 1 and the LCFS will give a highly variable system residence time. The relationship between the variances is independent of the form of the service distribution, $B(t)$, and depends only on the server utilisation.

8.6   Exercises

1.   Consider the batch arrival $M/G/1$ queue. Batches of jobs arrive, with the time between batches being exponentially distributed with λ. Each batch contains a geometrically distributed number of jobs.

2.   Evaluate the second moments of the delay busy period and the delay cycle, and show that they satisfy:

$$E[T_H^2] = \frac{\lambda E[S^2]}{(1-\rho)^3} E[T_I] + \frac{\rho^2}{(1-\rho)^2} E[T_I^2]$$

$$E[T_D^2] = \frac{\lambda E[S^2]}{(1-\rho)^3} E[T_I] + \frac{E[T_I^2]}{(1-\rho)^2}$$

3.   Analyse the $M/G/1$ queue with breakdowns and repairs of the server. Assume that operative intervals are exponentially distributed with parameter γ, and repair times are generally distributed with mean $\bar{R}$, second moment $E[R^2]$, and Laplace transform $R^*(s)$. If interrupted jobs resume processing at the point of interruption, then show that the elapsed time from the start of a job's service until its completion has Laplace transform:

$$A^*(s) = B^*(s + \gamma(1 - R^*(s)))$$

Jobs that arrive to find the server under repair must wait for the residual life of the repair before starting their service. Derive the Laplace transform of the time from arrival until completion of a job that arrives to an empty system. Hence derive the probability generating function of queue length, and confirm the mean number in the system as calculated in section 6.2.

8.7 Further reading

Takács discovered the elegant technique of delay cycle analysis[4]. It has been used by Gaver to analyse queues with breakdowns[2]. Its use to find waiting time distributions was first reported in the book by Conway, Maxwell, and Miller[1]. They provide an excellent description of the technique, on which this presentation is based. It has been applied to multiple server systems by Omahen and Marathe[3].

8.8 Bibliography

[1] R.W. Conway, W.L. Maxwell, and L.W. Miller. *Theory of Scheduling*. Addison-Wesley, Reading, Massachusetts, 1967.

[2] D.P. Gaver. A waiting line with interrupted service, including breakdowns. *Journal of the Royal Statistical Society (Series B)*, 24:73–90, 1962.

[3] K.J. Omahen and V. Marathe. Analysis and applications of the delay cycle for the $M/M/c$ queueing system. *Journal of the ACM*, 25(2):283–303, 1978.

[4] L. Takács. *Introduction to the Theory of Queues*. Oxford University Press, Oxford, 1962.

Chapter 9

Multiple server queues

In this chapter we examine the effects of increasing the number of servers available to process jobs. It seems obvious that adding extra servers will decrease the mean waiting time of jobs, but we shall see that this is not universally true. We also examine the trade off between adding an extra server, and replacing the current server with a more powerful one. A system with a second server available, but only after some delay is also treated.

9.1 $M/M/c$ queues

The $M/M/c$ system can be easily solved in the same way as the $M/M/1$ system. The arrival rate of jobs is λ, and each job requires an exponentially distributed service with mean $1/\mu$. The jobs join a common queue, and when a server becomes available, they are served. As before the state of the system is the number of jobs present. If there are fewer than c jobs present then they are all in service. In state i, the rate at which jobs leave the system is $\min(i, c)\mu$. Solution of the balance equations then leads to:

$$
\pi_i = \begin{cases} \frac{(c\rho)^i}{i!}\pi_0, & i = 0, 1, \ldots, c \\ \frac{c^c\rho^i}{c!}\pi_0, & i \geq c \end{cases} \tag{9.1}
$$

where $\rho = \lambda/(c\mu)$. This choice for ρ means that the condition for stability is still $\rho < 1$. It represents the amount of work brought into the system per unit time as a fraction of the total capacity of the system. The idle probability, π_0, is found from the normalisation condition.

$$
\pi_0 = \left[\sum_{i=0}^{c-1} \frac{(c\rho)^i}{i!} + \frac{c^c}{c!} \cdot \frac{\rho^c}{(1-\rho)} \right]^{-1} \tag{9.2}
$$

The mean queue length can be evaluated from the definition:

$$
\bar{N} = \sum_{i=1}^{\infty} i\pi_i
$$

First we separate the terms into those in which all customers present are being served, and those in which some customers are queueing.

$$
\begin{aligned}
\bar{N} &= \sum_{i=1}^{c} i\pi_i + \sum_{i=c+1}^{\infty} i\pi_i \\
&= \pi_0 \sum_{i=1}^{c} i\frac{(c\rho)^i}{i!} + \pi_c \sum_{i=c+1}^{\infty} i\rho^{i-c} \\
&= \pi_0 \sum_{i=1}^{c} i\frac{(c\rho)^i}{i!} + \pi_c\rho^{1-c} \sum_{i=c+1}^{\infty} i\rho^{i-1} \\
&= \pi_0 \sum_{i=1}^{c} i\frac{(c\rho)^i}{i!} + \pi_c\rho^{1-c} \sum_{i=c+1}^{\infty} \frac{\mathrm{d}}{\mathrm{d}\rho}\rho^{i-1} \\
&= \pi_0 \sum_{i=1}^{c} i\frac{(c\rho)^i}{i!} + \pi_c\rho^{1-c} \frac{\mathrm{d}}{\mathrm{d}\rho}\frac{\rho^{c+1}}{1-\rho} \\
&= \pi_0 \sum_{i=1}^{c} i\frac{(c\rho)^i}{i!} + \pi_c\rho^{1-c}\frac{(c+1)\rho^c(1-\rho)+\rho^{c+1}}{(1-\rho)^2} \\
&= \pi_0 \sum_{i=1}^{c} i\frac{(c\rho)^i}{i!} + \pi_c\frac{c\rho(1-\rho)}{(1-\rho)^2} + \pi_c\frac{\rho(1-\rho)}{(1-\rho)^2} + \pi_c\frac{\rho^2}{(1-\rho)^2} \\
&= c\rho\pi_0 \left(\sum_{i=0}^{c-1} \frac{(c\rho)^i}{i!} + \frac{(c\rho)^c}{c!}\frac{1}{(1-\rho)} \right) + \pi_c\frac{\rho(1-\rho)}{(1-\rho)^2} + \pi_c\frac{\rho^2}{(1-\rho)^2} \\
&= c\rho + \pi_c\frac{\rho}{(1-\rho)^2}
\end{aligned}
$$

This provides a convenient formula for computing the mean queue length and the mean delay. The mean number of customers which have to queue before entering service is given by $\sum_{i=c+1}^{\infty}(i-c)\pi_i$ and from the above analysis can be seen to equal $\pi_c\rho/((1-\rho)^2)$. Thus the mean number of customers in service, which equals the mean number of active servers is given by $c\rho = \lambda/\mu$. That this formula for the number of busy servers holds for the $G/G/c$ queue can be shown using Little's theorem. Consider the servers. The arrival rate to the servers is λ. (The presence of a queue will not affect the mean arrival rate so long as the system is stable.) The mean time in the system is $1/\mu$. Hence by Little's theorem, the mean number of jobs in the system, which is the mean number of active servers, equals the traffic intensity λ/μ.

It is interesting to consider the following three systems; (a) two $M/M/1$ systems with separate queues, in which an arrival chooses a queue at random; (b) an $M/M/2$ system, two identical servers serving a common queue; and (c) an $M/M/1$ system with a server of twice the processing rate of the servers in system (b). Notice that the maximum rate of processing jobs is the same in all three systems. The mean number of jobs in system (a) is just twice the mean number of jobs in a single $M/M/1$ queue. In system (c), the mean is given by the formula for an $M/M/1$ queue with traffic intensity $\rho/2$. Substituting 2 for c in equation(9.1), we find that $\pi_1 = \rho\pi_0$, $\pi_2 = (\rho^2/2)\pi_0$, and, in general, $\pi_j = 2(\rho/2)^j\pi_0$ for $j \geq 2$. π_0 is found from equation (9.2). System (c) has the shortest queue, then system (b), and system (a) the largest number of jobs waiting. The explanation of this result is found *not* by comparing the maximum processing rate of the systems, which are identical, but in the processing rates when the number of jobs in the system is small. System (c) always serves at full speed until there are no jobs waiting; system (b) serves at the same rate unless there is only one job in the system, when it serves at half the usual rate; whilst system (a) can have its rate reduced to half when one of the queues empties, even though the other may have a large number of jobs waiting.

9.2 $M/M/\infty$ queueing system

The idea of an infinite number of servers seems like a mathematician's dream. In some sense it is, but the analysis of such a system is important for two reasons. First, it gives bounds on the performance which it is possible to attain merely by increasing the number of servers. Secondly, it can be used to model the effect of delay in larger systems. If the service time is used to represent the time taken to move from one queue to another, then the mean number in the $M/M/\infty$ queue is the mean number of messages in transit.

As usual, we assume Poisson arrivals at rate λ, and exponentially distributed service requests with mean $1/\mu$. Since there are as many servers as there are customers in the system, the rate of leaving state i for state $i - 1$ is $i\mu$. The balance equations are equally simple:

$$\lambda\pi_i = (i + 1)\mu\pi_{i+1}$$

which can be easily solved for π_i to give:

$$\pi_i = \frac{\rho^i}{i!}\pi_0$$

The normalisation condition gives $\pi_0 = e^{-\rho}$. Notice that there are no restrictions on the values of λ or μ. Since every customer receives service independently, any number can be dealt with simultaneously. The generating function of the number in the system is $G(z) = e^{\rho(z-1)}$ and the mean number of customers is ρ. Notice that we could have derived the mean number in the system using Little's theorem without having to derive the state probabilities; the mean time spent in the system is $1/\mu$ and the mean arrival rate is λ. Thus the mean number in a $G/G/\infty$ system will be ρ, independent of the distributions involved.

9.3   Slow servers

The analyses that we have given in the previous sections relate to multiserver systems in which the servers are all identical. When the servers are not identical, a new problem is evident: is it always optimal to keep all servers active? Alternatively, are there situations when the performance of the system will be improved by allowing a server to stand idle? Although it seems that it is a good idea to keep all servers active for as long as possible, and not to allow any to remain idle whilst there is work to be done, this is not the optimal solution if the waiting times are to be minimised. Consider a system with two servers, server 1 being much faster than server 2. A customer arriving to find only a single customer in the system, being served by server 1, could find if he was allocated to server 2, that server 1 finished service soon after, and the customer is condemned to a long service by server 2, instead of a fast service by server 1.

Let us consider this situation in more detail, and analyse the situation of two exponential servers of different service rates serving a Poisson stream of arrivals. The state of the system can be given by the number of jobs in the system, except when there is a single job present. In that case, we must distinguish between the case of a single job which is being served by the fast server and being served by the slow server. Assume that arrivals are at rate λ, and the two servers have processing rates μ_1 and μ_2. Without loss of generality, we can assume that $\mu_1 > \mu_2$. We use μ to stand for the service rate when both servers are active i.e., $\mu_1 + \mu_2$. Denote the steady state probability of there being i jobs in the system by π_i, with the case of only a single job in the system being denoted by the two states 1,1 and 1,2, representing server 1 and server 2, respectively, being the active server. There is a further aspect of this problem that we must specify before the balance equations can be constructed. If the arriving customers are aware of the distinction between the two servers, then a newly arrived customer, finding both servers idle, would always choose server 1. If the customers are ignorant of the distinction between the servers then either will be chosen with equal probability.

We shall first analyse the 'no knowledge' situation, and then investigate the case when the customers know which server is faster. The balance equations are easily stated:

$$\lambda \pi_0 = \mu_1 \pi_{1,1} + \mu_2 \pi_{1,2} \tag{9.3}$$

$$(\lambda + \mu_1)\pi_{1,1} = \lambda/2\pi_0 + \mu_2 \pi_2 \tag{9.4}$$

$$(\lambda + \mu_2)\pi_{1,2} = \lambda/2\pi_0 + \mu_1 \pi_2 \tag{9.5}$$

$$\lambda \pi_2 = \mu \pi_3 \tag{9.6}$$

$$\lambda \pi_i = \mu \pi_{i+1} \tag{9.7}$$

If we write $\pi_1 = \pi_{1,1} + \pi_{1,2}$, and $\rho = \lambda/\mu$, then it easily follows that:

$$\pi_j = \rho^{j-1}\pi_1, \quad j \geq 2$$

Substituting $\rho(\pi_{1,1} + \pi_{1,2})$ for π_2 in equations (9.4) and (9.5), we can solve in terms of π_0 so that:

$$\pi_{1,1} = \frac{\lambda\pi_0}{2\mu_1}$$

$$\pi_{1,2} = \frac{\lambda\pi_0}{2\mu_2}$$

Hence:

$$\pi_1 = \frac{\lambda\pi_0}{a}$$

where $a = 2\mu_1\mu_2/\mu$. Hence:

$$\pi_0 = \frac{1 - \rho}{1 - \rho + \lambda/a}$$

The mean queue length can be found by calculating the generating function and differentiating at 1.

$$G(z) = \sum_{i=0}^{\infty} \pi_i z^i$$

$$= \pi_0 + \frac{\pi_1 z}{1 - \rho z}$$

$$= \pi_0 \frac{1 - \rho z + (\lambda/a)z}{1 - \rho z}$$

Hence:

$$\bar{N} = G'(1) = \pi_0 \frac{\lambda/a}{(1 - \rho)^2}$$

$$= \frac{\lambda}{(1 - \rho)(\lambda + (1 - \rho)a)} \tag{9.8}$$

This is to be compared with the mean queue length which would result in an $M/M/1$ system with only a server of rate μ_1. The two server system is better only if:

$$\bar{N} < \frac{\lambda}{\mu_1 - \lambda}$$

or:

$$\mu_1 - \lambda < (1 - \rho)(\lambda + (1 - \rho)a)$$

Rearranging this expression, the slow server is only worth having if:

$$\rho^2(\mu - a) - 2\rho(\mu - a) - (a - \mu_1) < 0$$

The behaviour of the left hand side can be investigated as a quadratic equation. The points of interest are where it changes sign, i.e. its roots. This quadratic expression in ρ has a single root, ρ_c, in the range $0 \leq \rho \leq 1$. This is easily seen since the sign of the expression is negative at 1, and positive at 0. When ρ is less than this root, the second server contributes nothing to the performance of the system. The root ρ_c is:

$$\rho_c = 1 - \sqrt{\frac{\mu_2}{\mu - a}}$$

It is convenient to express this root in terms of r, the ratio of the service rates.

$$r \triangleq \frac{\mu_2}{\mu_1}$$

$$\rho_c = 1 - \sqrt{\frac{r(1 + r)}{1 + r^2}}$$

This gives a criterion for use of the second server based on r, which is dimensionless: if the overall traffic intensity in a two server system is less than ρ_c then the mean delay will be improved if the slower server is not used. The system with the slow server removed will be stable if $\rho < \rho_c$ since:

$$\lambda/\mu < \rho_c$$
$$\lambda/\mu_1 < \rho_c \mu / \mu_1$$
$$= (1 + r)\rho_c$$
$$= (1 + r) - \sqrt{\frac{(1 + r)^3 r}{1 + r^2}}$$
$$= (1 + r) - \sqrt{\frac{r^4 + 3r^3 + 3r^2 + r}{1 + r^2}}$$
$$= (1 + r) - \sqrt{r^2 + \frac{3r^3 + 2r^2 + r}{r^2 + 1}}$$
$$< 1$$

If customers know which is the fast server, then obviously they will choose that server whenever they have the option. In this case, a similar analysis can be made. The balance equations are identical to those in the uninformed case except for equations (9.4) and (9.5) which are replaced by:

$$(\lambda + \mu_1)\pi_{1,1} = \lambda\pi_0 + \mu_2\pi_2$$
$$(\lambda + \mu_2)\pi_{1,2} = \mu_1\pi_2$$

As before, we can solve the balance equations and find:

$$\pi_{1,1} = \frac{\lambda(\lambda + \mu)\pi_0}{\mu_1(2\lambda + \mu)}$$

$$\pi_{1,2} = \frac{\lambda^2 \pi_0}{\mu_2(2\lambda + \mu)}$$

$$\pi_1 = \frac{\lambda \pi_0}{a^*}$$

with:

$$a^* = \frac{(2\lambda + \mu)\mu_1\mu_2}{\mu(\lambda + \mu_2)}$$

As before,

$$\pi_0 = \frac{1 - \rho}{1 - \rho + \lambda/a^*}$$

and the mean number in the system is:

$$\bar{N} = \frac{\lambda}{(1 - \rho)(\lambda + (1 - \rho)a^*)}$$

Although superficially identical to equation (9.8), there is one major difference, a^* is a function of λ as well as μ_1 and μ_2, so the dimensionless analysis becomes more involved. As before, it is worth having the slow server if $\bar{N} < \lambda/(\mu_1 - \lambda)$ which is:

$$\rho^2(\mu - a^*) - 2\rho(\mu - a^*) - (a^* - \mu_1) < 0$$

This can be simplified by observing that $\lambda = \mu \cdot \rho$, so that $a^* = \mu_1\mu_2(1 + 2\rho)/(\rho\mu + \mu_2)$. Substituting, and taking over the common denominator $\rho\mu + \mu_2$, which is positive for all ρ in the range $0 \le \rho \le 1$, it follows that:

$$\rho^2(\rho(\mu_1^2 + \mu_2^2) + \mu_2^2) - 2\rho(\rho(\mu_1^2 + \mu_2^2) + \mu_2^2) + \rho\mu_1(\mu_1 - \mu_2) < 0$$

As before, we are interested in the roots of the left hand side. It is a cubic, but ρ is a common factor, so we have a root $\rho = 0$, and a quadratic expression which can be rewritten in terms of $r = \mu_2/\mu_1$, as:

$$\rho^2(1 + r^2) - \rho(2 + r^2) - (2r - 1)(1 + r) \tag{9.9}$$

The roots of (9.9) give us the information needed to decide when to remove the slow server. Label the smaller root as ρ_0 and the larger one as ρ_0^*. As the ratio of the speeds of the two servers, r, changes, so does the condition under which the slow server is needed. The larger root, ρ_0^* is always greater than 1, so it has no influence on the decision, since only values of ρ in the range $0 \le \rho < 1$ can lead to a stable system. When $r = 0$, both roots are 1, so we will never need the slow server, which is as we expect. As r increases, the larger root increases, and the smaller one decreases. At $r = 0.5$, $\rho_0 = 0$ and for larger r is negative. Thus the slow server is always needed when $r > 0.5$, whatever the traffic intensity. When the fast server is twice as fast as, or faster than, the slow server, the slow server should be removed if the traffic intensity is less than ρ_0. It can be shown that the single server system obtained by removing the slow server will be stable if this condition is satisfied.

It is possible to analyse this problem in the case where the jobs have unreliable knowledge as to which server is faster. If jobs choose the faster server with probability p, then the factor $1/2$ in equations (9.4) and (9.5) is replaced by p and $1 - p$, respectively. The analysis proceeds as before, and expressions result identical to the case of informed customers for the idle probability, π_0, and the mean queue length. The difference is that a^* is a function of p. The analysis of the conditions under which the slow server should be removed is more involved. The cases treated above of fully informed customers corresponds to $p = 1$ and that of uninformed customers corresponds to $p = 0$.

9.4   System with delayed second server

Some systems allow extra service capacity to be requested. If extra capacity can be made available immediately, the problem is a straightforward application of Markovian queueing analysis. However in some systems, there is a delay before the extra capacity becomes available. An example might be found in the transport layer protocols of a communications network. Multiple transport connections can be multiplexed over a single network connection, which is able to provide a certain level of throughput. If the transport connections provide too much traffic for an interval, the throughput will degrade, so the transport protocol can open a second network connection, and spilt the traffic between the two network connections. Opening a network connection is a non-trivial operation and takes a significant time. Since opening connections costs money, and increased delay also costs money, the designers of the protocol need to know the trade-off between requesting a second connection frequently, and not requesting one and suffering an increased delay to messages. We model this system using an $M/M/2$ system. One server is available permanently. Occasionally, the number of jobs waiting in the queue becomes too large, and to avoid excessive delay, a second server is requested as soon as the queue length has reached K. This second server is not immediately available, but only after a (random) delay. The request for the second server is *not* cancelled if the queue length falls below K, but only if the queue becomes empty. We shall ascertain the queue occupancy and determine the effect of the threshold K on the operating cost of the queue.

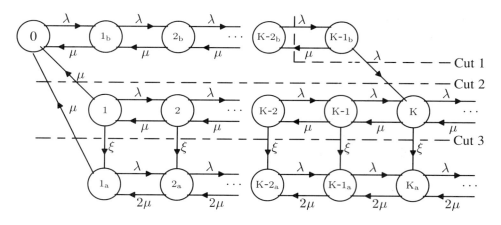

Figure 9.1 Queueing system with delayed second server

We assume that jobs arrive at the queue according to Poisson process of rate λ. The lengths of the jobs and the time to make the reserve server available are exponentially distributed with means $1/\mu$ and $1/\xi$, respectively. The servers have equal transmission capacity; thus, they can be represented as servers of equal rates. The reserve server is released instantly when the queue is empty.

This system can be modeled as a Markovian queueing system with state diagram shown in Figure 9.1. The upper branch (states $i_b, 0 \leq i < K$) represents the queue before the reserve server is requested, the states in the middle branch (states $i, 1 \leq i \leq \infty$) represent the queue after the request but before the second server is available, and finally, the states in the lowest branch (states $i_a, 1 \leq i \leq \infty$) represent the queue after the second server has been made available, when it is being served by both servers. The steady state probabilities of the states in the upper, middle and lower branches are denoted as q_i, p_i, and r_i, respectively.

To determine the steady state probabilities, we begin by considering cuts of the type marked by dashed line (1) in Figure 9.1 that yield the following balance equations for the upper branch:

$$\lambda q_{K-i} = \mu q_{K-i+1} + \lambda q_{K-1}, \quad 2 \leq i \leq K$$

that allows us to express the probabilities of states in the upper branch as a function of the probability of state $(K-1)_b$:

$$q_{K-i} = q_{K-1} \sum_{j=0}^{i-1} (\mu/\lambda)^j, \quad 1 \leq i \leq K \tag{9.10}$$

whose sum is given by:

$$\sum_{i=0}^{K-1} q_i = \frac{q_{K-1}}{1 - (\mu/\lambda)} \left[K - (\mu/\lambda) \frac{1 - (\mu/\lambda)^K}{1 - (\mu/\lambda)} \right] \tag{9.11}$$

the following two useful balance equations result from the cuts 2 and 3 in Figure 9.1:

$$(p_1 + r_1)\mu = \lambda q_{K-1} \tag{9.12}$$

$$\xi \sum_{i=1}^{\infty} p_i = \mu r_1 \tag{9.13}$$

balance equations using cuts similar to cut 2, we derive a set of balance equations that relate q_{K-1} with p_i and r_i, $1 \le i \le \infty$. By summing these equations we get:

$$\mu \sum_{i=1}^{\infty} p_i + 2\mu \sum_{i=1}^{\infty} r_i = \lambda \sum_{i=1}^{\infty} p_i + \lambda \sum_{i=1}^{\infty} r_i + \mu r_1 + K\lambda q_{K-1} \tag{9.14}$$

Combining equations (9.13) and (9.14) yields the following expression for the probability that both servers are in use, $\sum_{i=1}^{\infty} r_i$:

$$\sum_{i=1}^{\infty} r_i = \frac{[\mu - \mu(\mu - \lambda)/\xi]r_1 + \lambda K q_{K-1}}{2\mu - \lambda} \tag{9.15}$$

From the normalisation condition:

$$\sum_{i=0}^{K-1} q_i + \sum_{i=1}^{\infty} p_i + \sum_{i=1}^{\infty} r_i = 1$$

and using equations (9.11), (9.13), and (9.15), we get the following equation relating r_1 to q_{K-1}:

$$\frac{\mu(\xi + \mu)}{\xi(2\mu - \lambda)} r_1 + \left[\frac{\lambda K}{2\mu - \lambda} + \frac{K(1 - \mu/\lambda) - (\mu/\lambda)\left[1 - (\mu/\lambda)^K\right]}{(1 - (\mu/\lambda))^2} \right] q_{K-1} = 1$$

To solve for r_1 and q_{K-1}, we need one more equation, which is obtained by considering the generating function of the middle branch $P(z) = \sum_{i=1}^{\infty} p_i z^i$. This function can be derived from the set of balance equations, each of which isolates a single state in this branch:

$$P(z) = \frac{\lambda z^K q_{K-1} - \mu p_1}{\xi + \mu(1 - z^{-1}) + \lambda(1 - z)} \tag{9.16}$$

The denominator has two real roots that are both positive. Unless the numerator is also zero at these points, the generating function will have a pole there. From its definition as a power series, $P(z)$ can have no poles for $|z| < C$, the radius of convergence. Because $P(z)$ is a probability generating function $C \ge 1$. The smaller root:

$$z_p = \frac{(\lambda + \mu + \xi) - \sqrt{(\lambda + \mu + \xi)^2 - 4\lambda\mu}}{2\lambda} \qquad (9.17)$$

is always inside the unit circle. In this region, the numerator of the generating function must have a zero at z_p, which leads to the following equation:

$$\lambda z_p^K q_{K-1} - \mu p_1 = 0 \qquad (9.18)$$

Using equations (9.12), we get the second equation relating r_1 and q_{K-1}:

$$\mu r_1 = \lambda q_{K-1} \left[1 - z_p^K \right]$$

We proceed now to evaluate the average number in the queue which is given by:

$$\bar{N} = \sum_{i=1}^{K-1} i q_i + \sum_{i=1}^{\infty} i p_i + \sum_{i=1}^{\infty} i r_i = \sum_{i=1}^{K-1} i q_i + P'(1) + R'(1)$$

Using equations (9.10), we can express the first term by:

$$\sum_{i=0}^{K-1} i q_i = \frac{q_{K-1}}{1 - \mu/\lambda} \left[\frac{K(K-1)}{2} - \frac{\mu/\lambda}{1 - \mu/\lambda} \cdot \frac{(\mu/\lambda)^K - K\mu/\lambda + (K-1)}{1 - \mu/\lambda} \right]$$

The generating function of the lower branch, $R(z) = \sum_{i=1}^{\infty} r_i z^i$ can be obtained by the set of balance equations each of which isolates one i_a state and is expressed by:

$$R(z) = \frac{\xi P(z) + \mu(z - 2)r_1}{\lambda(1 - z) - 2\mu(1 - z^{-1})}$$

Evaluating its derivative at $z = 1$ yields the following expression, after application of l'Hospital's rule:

$$R'(1) = \frac{(2\mu - \lambda)\xi P''(1) + 4\mu(\xi P'(1) + \mu r_1)}{2(2\mu - \lambda)^2}$$

where:

$$P'(1) = \frac{K\xi - (\mu - \lambda)(1 - z_p^K)}{\xi^2} \lambda q_{K-1}$$

$$P''(1) = \left\{ \xi[\xi K(K-1) + 2\mu(1 - z_p^K)] \right.$$
$$\left. - 2(\mu - \lambda)[K\xi - (1 - z_p^K)(\mu - \lambda)] \right\} \frac{\lambda q_{K-1}}{\xi^3}$$

are obtained by differentiating equations (9.16).

The rate at which the queue *releases* the reserve server is given by μr_1, which is exactly the rate at which that server is connected to the queue. Applying costs to the job waiting time (c_1) and to each initialisation of the second server (c_2), and using Little's result, the average cost, (C) to the system is given by:

$$C = c_1 \frac{\bar{N}}{\lambda} + c_2 \mu r_1$$

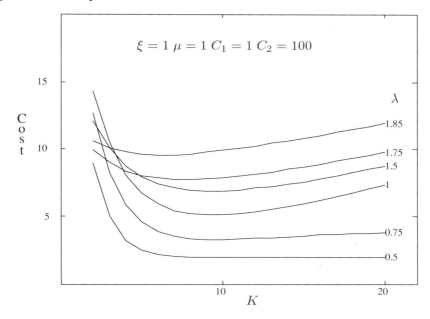

$\xi = 1 \; \mu = 1 \; C_1 = 1 \; C_2 = 100$

Figure 9.2 Cost as a function of threshold, slow access

9.4.1 Numerical results

Numerical results of the model are presented in Figures 9.2, 9.3, and 9.4. Figures 9.2 and 9.3 depict the overall cost of operating the queue as a function of the threshold. The normalised rate λ/μ is the independent variable. Figure 9.2 shows the cost when ξ, the rate at which the reserve server becomes available, is low. In Figure 9.3 the rate is relatively high.

As Figures 9.2 and 9.3 show, at light load $(\lambda/\mu < 1)$ the cost curve monotonically decreases with K. The high cost at low values for K is because, at low loads, the probability of high queue occupancy is low. The delay, then, is quite small even with only one server, and the second server, when called, will be of little help because it may find only one message in the queue that is already being served by the permanent server. Moreover, under light load the expected busy period is short, which means that the queue moves quite frequently into and out of the zero state. A low threshold means that the reserve server is called frequently, thus implying high initialisation costs. Because initialisation costs are incurred only when the initialisation is completed to actually connect the server, the higher the initialisation speed, as measured by ξ, the more often the server is connected before the queue is empty. This explains the higher cost at low load and low threshold shown in Figures 9.2 compared to 9.3, which represent large and small ξ, respectively.

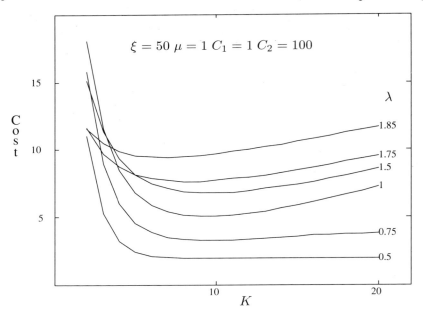

Figure 9.3 Cost as a function of threshold, fast access

When the arrival rate is close to capacity $(\lambda/\mu \approx 2)$, there is little probability that the second server finds the queue empty when it is connected. Thus, that server is expected to be of real help, which means that the sooner it is called, the lower the expected delay will be which is, in this case, the dominant factor in cost. For this reason, the cost is an increasing function of K. In a moderately loaded system, $(\lambda/\mu \approx 1.5)$, the two aforementioned effects combine to produce a cost curve that has a minimum at some threshold K, which is then the optimum threshold. Note, however, that the slope of the curve near the minimum is small, which implies that working not exactly with the optimum threshold but somewhere in its vicinity does not increase the cost very much.

Figure 9.4 depicts the cost as a function of the load when K is fixed, and we observe that the cost increases with the load for large K but has a minimum for small K. Because the reserve server is called only when needed, the increase in cost is due to the increase in delay. However, for small K the reserve server is called too often at light load, which contributes to the high cost. As the load increases, the rate at which the reserve server is called decreases while the delay increases only moderately. The overall effect, however, is to decrease the total cost.

At heavy loads, the threshold has little effect on the cost because, at close to capacity, the two servers are needed most of the time. Busy periods are long,

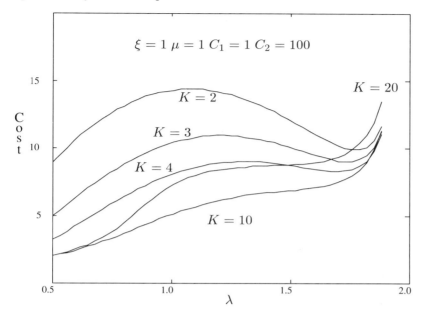

$\xi = 1\ \mu = 1\ C_1 = 1\ C_2 = 100$

Figure 9.4 Cost as a function of load, slow access

which implies low cost due to the infrequency of requesting the second server, thus the buffering delay, which is relatively insensitive to K at high load, determines the overall cost. This also explains the sharp increase in cost at this load region.

9.5   Multiprocessor systems with priorities

We have examined the effect of giving one class of jobs priority over another in single processor systems. We now extend this to a study of multiprocessor systems. The main difference is that when we treat a single processor system, the state of the system is very simple to define. In a multiprocessor system the state space description can become very involved, and the state space very large, because potentially each processor can be serving a different class of job.

We consider only the simplest case, where N identical processors are giving service to jobs from two different classes. Jobs from class 1 have preemptive resume priority over jobs from class 2. If we assume that arrivals in class i form a Poisson stream with mean rate λ_i, and that the service requirement of a class i job is exponentially distributed with mean $1/\mu_i$, then the system is Markovian, and the state can be represented by a pair of integers, the number of jobs of each class present. The system will reach steady state if the amount of work brought into the system in unit time is less than the amount that can be served, that is,

$$\frac{\lambda_1}{\mu_1} + \frac{\lambda_2}{\mu_2} < N$$

Assuming that the system reaches steady state, we define the probability of state (i, j) by π_{ij}. The balance equations can be simply stated, writing $\sigma_c(i, j)$ for the instantaneous departure rate of class c jobs when the system state is (i, j).

$$\pi_{ij}[\lambda_1 + \lambda_2 + \sigma_1(i,j) + \sigma_2(i,j)] = \lambda_1\pi_{i-1,j} + \lambda_2\pi_{i,j-1}$$
$$+ \sigma_1(i+1,j)\pi_{i+1,j}$$
$$+ \sigma_2(i,j+1)\pi_{i,j+1}$$

Conventionally, we set $\pi_{i,-1} = \pi_{-1,j} = 0$. Now the instantaneous completion rates in state (i, j) can be written:

$$\sigma_1(i,j) = \mu_1 \cdot \min(i, N)$$
$$\sigma_2(i,j) = \mu_2 \cdot \min(j, N - i)$$

As far as class 1 jobs are concerned, the system appears as an $M/M/N$ system, since they have preemptive priority over class 2 jobs.. The marginal queue length probabilities for class 1 jobs are thus given by the formulas calculated in section 9.1.

A complete solution is more complex. We consider the balance equations in four non-intersecting subsets;

1. $0 \le i, j \le N - 1$;
2. $i \ge N, 0 \le j \le N - 1$;
3. $j \ge N, 0 \le i \le N - 1$;
4. $i, j \ge N$.

In subset 1, the transition rates depend on both i and j. In subsets 2 and 4, they are state independent, and in subset 3, they depend on the value of i only. These subsets are marked on the state space diagram, Figure 9.5.

Considering subset 1, we find N^2 equations, involving $N^2 + 2N$ unknown probabilities. These unknown probabilities are π_{ij} for $i, j = 0, 1, \dots, N - 1$, $\pi_{N,j}$ for $j = 0, 1, \dots, N - 1$, and $\pi_{i,N}$ for $i = 0, 1, \dots, N - 1$.

In subset 2, the transition rates are state independent, (in fact $\sigma_1 = N\mu_1$, and $\sigma_2(i,j) = 0$.) We define the generating functions $g_j(u) = \sum_{i=N}^{\infty} \pi_{ij} u^{i-N}$. Taking the balance equation, specialised to states in subset 2, we have:

$$\pi_{ij}[\lambda_1 + \lambda_2 + N\mu_1] = \lambda_1\pi_{i-1,j} + \lambda_2\pi_{i,j-1} + N\mu_1\pi_{i+1,j}$$

Multiplying each of these equations, $(i = N, \dots)$, by u^{i-N+1} and summing over $i \ge N$:

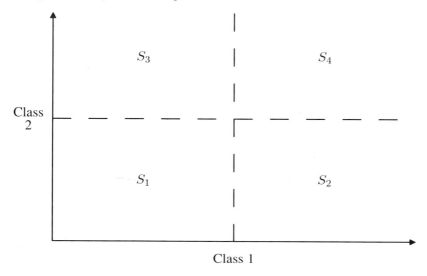

Figure 9.5 State space for $M/M/N$ preemptive priority system

$$a(u)g_j(u) = \lambda_2 g_{j-1}(u) + \lambda_1 u \pi_{N-1,j} - N\mu_1 \pi_{N,j} \qquad (9.19)$$

where:

$$a(u) = \lambda_1 u(1 - u) + \lambda_2 u - N\mu_1(1 - u)$$

Since $g_{-1}(u) \equiv 0$ by definition, we can easily solve these relations, and discover that:

$$g_j(u) = \left(\sum_{k=0}^{j} [a(u)]^k [\lambda_1 u \pi_{N-1,k} - N\mu_1 \pi_{N,k}](\lambda_2)^{j-k} \right) \Big/ \left([a(u)]^{j+1} \right) \quad (9.20)$$

All the probabilities that appear in these equations for $j = 0, 1, \ldots, N - 1$ have already appeared in the equations for subset 1. Now, $a(u)$ is a quadratic in u which has a single real root $u = u^\star$ such that $0 < u^\star < 1$. This is easily seen since $a(0) < 0$ and $a(1) > 0$. Thus the denominator in the equation for g_{N-1} has a zero of order N at $u^\star$. But g_{N-1} is a power series with radius of convergence at least 1, so there can be no singularities at $u^\star$. Therefore the numerator must have a zero of similar order at $u^\star$. This is exactly equivalent to the numerator and its first $N - 1$ derivatives having zeros at $u^\star$. This gives a further N equations.

In subset 3, the departure rates depend on i, but not on j. Again we define the generating functions $\tilde{g}_i(v) = \sum_{j=N}^{\infty} \pi_{ij} v^{j-N}$. As before, we can reduce the balance equations to a set of equations relating the generating functions $\tilde{g}_i$ for $i = 0, 1, \ldots, N - 1$ and the unknown probabilities $\pi_{i,N-1}$ and $\pi_{i,N}$ for $i = 0, 1, \ldots, N$.

$$a_i(v)\tilde{g}_i(v) = \lambda_1 v \tilde{g}_{i-1}(v) + (i+1)\mu_1 v \tilde{g}_{i+1}(v) \tag{9.21}$$
$$+ \lambda_2 v \pi_{i,N-1} - (N - i)\mu_2 \pi_{i,N}$$

where:

$$a_i(v) = \lambda_1 v + \lambda_2 v(1 - v) + i\mu_1 v - (N - i)\mu_2(1 - v)$$

These equations can be solved to express each $\tilde{g}_i(v)$ in terms of $\tilde{g}_0(v)$ and the unknown probabilities, $\pi_{i,N-1}$ and $\pi_{i,N}$.

Writing equations (9.21) in matrix notation,

$$A(v)\tilde{\mathbf{g}}(v) = \mathbf{c}(v)\tilde{g}_0(v) + \mathbf{b}(v) \tag{9.22}$$

with:

$$A(v) = \begin{bmatrix} -\mu_1 v & 0 & 0 & \cdots & \cdots & 0 \\ a_1(v) & -2\mu_1 v & 0 & \cdots & \cdots & 0 \\ -\lambda_1 v & a_2(v) & -3\mu_1 v & \cdots & \cdots & 0 \\ \vdots & \vdots & \vdots & \vdots & \vdots & \vdots \\ 0 & 0 & \cdots & -\lambda_1 v & a_{N-1}(v) & -N\mu_1 v \end{bmatrix}$$

$$\tilde{\mathbf{g}}(v) = [\tilde{g}_1(v), \tilde{g}_2(v), \ldots, \tilde{g}_N(v)]$$

$$\mathbf{c}(v) = [-a_0(v), \lambda_1 v, 0, \ldots, 0]$$

$$\mathbf{b}(v) = [(\lambda_1 v \pi_{0,N-1} - N\mu_2 \pi_{0,N}), \ldots, (\lambda_1 v \pi_{N-1,N-1} - \mu_2 \pi_{N-1,N})]$$

We now apply Cramer's rule to solve the equations, and write:

$$\tilde{g}_{i(v)} = \frac{|C^i(v)|}{|A(v)|}\tilde{g}_0(v) + \frac{|B^i(v)|}{|A(v)|}, \quad i = 1, 2, \ldots, N \tag{9.23}$$

where $C^i(v)$ and $B^i(v)$ are the matrix $A(v)$ with its ith column replaced by $\mathbf{c}(v)$ and $\mathbf{b}(v)$, respectively; $|A|$ denotes the determinant of the matrix A. A number of simplifications are now possible, particularly because $A(v)$ is lower triangular, so its determinant is:

$$|A(v)| = (-1)^N N!(\mu_1 v)^N$$

$|B^i(v)|$ is a known function, and $|C^i(v)|$ is a known function which contains the unknown probabilities $\pi_{i,N-1}$ and $\pi_{i,N}$. Now $|B^i(v)|$ and $|C^i(v)|$ both contain factors of the form $[-(i+1)\mu_1 v] \cdots [-N\mu_1 v]$ (because of the structure of the determinants involved). These terms can effectively be factored out by considering only the principal minors of order i of each matrix, and we have:

$$\tilde{g}_{i(v)} = \frac{e_i(v)\tilde{g}_0(v) + f_i(v)}{v^i}, \quad i = 1, 2, \ldots, N \tag{9.24}$$

We can express all the functions $\tilde{g}_1, \tilde{g}_2, \ldots, \tilde{g}_N$ in terms of one unknown function, $\tilde{g}_0$, and the unknown probabilities, $\pi_{i,N-1}$ and $\pi_{i,N}$ for $i = 0, 1, 2, \ldots, N - 1$.

We now have $N^2 + 2N$ unknown probabilities, $N^2 + N$ equations relating them, and one unknown function $\tilde{g}_0(v)$. In subset 4, as in subset 2, the instantaneous departure rates are state independent. We can use the same definition of $g_j(u)$ in the subset $j \geq N$ as we use for $0 \leq j < N$. We introduce the two variable generating function:

$$G(u, v) = \sum_{i=N}^{\infty} \sum_{j=N}^{\infty} \pi_{i,j} u^{i-N} v^{j-N} = \sum_{j=N}^{\infty} g_j(u) v^{j-N}$$

Multiplying each of equations (9.19) by v^{j-N} and summing over all $j \geq N$ gives:

$$Q(u, v)G(u, v) = \lambda_2 u g_{N-1}(u) + \lambda_1 v \tilde{g}_{N-1}(v) - N\mu_1 \tilde{g}_N(v) \tag{9.25}$$

where:

$$Q(u, v) = \lambda_1 u(1 - u) + \lambda_2 u(1 - v) - N\mu_1(1 - u)$$

Now $g_{N-1}(u)$ is a known function given by equation (9.20) and $\tilde{g}_{N-1}(v)$ and $\tilde{g}_N(v)$ are both expressed in terms of $\tilde{g}_0(v)$ using equation (9.24). In the range $0 \leq v \leq 1$, the equation:

$$Q(u, v) = 0$$

has a single root $u = \eta(v)$ which also satisfies $0 < \eta(v) < 1$.

$$\eta(v) = \left[\lambda_1 + \lambda_2(1 - v) + N\mu_1 \right.$$
$$\left. - \sqrt{(\lambda_1 + \lambda_2(1 - v) + N\mu_1)^2 - 4N\lambda_1\mu_1} \right] / (2\lambda_1)$$

$Q(\eta(v), v) = 0$ and $G(\eta(v), v)$ is finite (for $|v| \leq 1$), so the right hand side of equation (9.25) must also equal 0 at all points $(\eta(v), v), |v| \leq 1$. Substitution of equation (9.24) gives an expression for:

$$\tilde{g}_0(v) = \left[-\lambda_2 \eta(v) v^N g_{N-1}(\eta(v)) - \lambda_1 \eta(v) v f_{N-1}(v) \right.$$
$$\left. + N\mu_1 f_N(v) \right] \tag{9.26}$$
$$/ \left[\lambda_1 \eta(v) v e_{N-1}(v) - N\mu_1 e_N(v) \right]$$

Now $\tilde{g}_0(v)$ must be finite for all $v, |v| \leq 1$. It is possible to show that the denominator in equation (9.26) has $N - 1$ distinct real roots in this region. At each of these roots the numerator must also equal 0 or $\tilde{g}_0$ would have a pole. Thus we have a further $N - 1$ equations all homogeneous.

The final equation is found using the normalisation condition. This can be expressed in several equivalent ways, for example:

$$1 - \rho_1 = \tilde{g}_0(1) + \sum_{j=0}^{N-1} \pi_{0j}$$

9.6   Exercises

1. Investigate the economics of adding extra servers to an $M/M/c$ queueing system. Assume that the system is stable with c servers. If customer queueing time costs c_q per customer per second, and an additional processor costs c_p per second, what are the conditions under which adding a further processor is economic?

2. Compare the $M/M/\infty$ system investigated in this chapter with the $M/M/1$ system with discouraged arrivals of chapter 4, exercise 2. What is the difference between them from the customers point of view? Which system gives a better mean delay?

3. Consider the delayed second server of section 9.4 when the threshold is 1. The steady state diagram no longer has it's upper branch. Evaluate the generating function of the number in the system.

4. Investigate a queueing system in which additional servers become available without any delay when the queue length reaches a threshold, k. The standard $M/M/2$ system is such a system with $k = 2$. It makes the second server available as soon as the second customer arrives in the system. When the queue length falls below the threshold again, there are two alternative policies. Either the second server can be dropped immediately, or it can be retained until the system is idle. The model investigated in this chapter is the second alternative, except for the finite request rate for the second server. Assume costs for requesting and/or using servers, and for customer delay, and compare the different policies.

9.7   Further reading

Markovian multiserver queues have been studied since the time of Erlang. The analysis of the $M/M/c$ system is found in many books, but the closed form for the mean queue length is only mentioned in Bhat[1].

Slow servers have an interesting history. The first reference to systems with heterogeneous servers appears to be Krishnamoorti[4] who solves the system assuming that the customers know the identity of the faster server. Lin and Kumar[5] analyse the problem from a control theory point of view, and evaluate optimal thresholds. Our presentation follows Rubinovitch[7, 8] who seems to be the first to analyse the system allowing for different levels of customer knowledge.

Systems with delay before the second server becomes available are introduced by King and Shacham[3]. Multiserver priority systems were solved exactly by Mitrani and King[6]. An alternative approximate method has been developed by Buzen and Bondi[2].

9.8 Bibliography

[1] U.N. Bhat. *Elements of Applied Stochastic Processes*. Wiley, New York, NY, 1984.

[2] J.P. Buzen and A.B. Bondi. The response time of priority classes under preemptive resume in $M/M/m$ queues. *Operations Research*, 31(3):456–465, May/June 1983.

[3] P.J.B. King and N. Shacham. Queueing analysis for buffers with dial-up servers. *Performance Evaluation*, 10(2):129–145, November 1989.

[4] B. Krishnamoorthi. On the Poisson queue with two heterogeneous servers. *Operations Research*, 11(3):321–330, May/June 1963.

[5] W. Lin and P.R. Kumar. Optimal control of a queueing system with two heterogeneous servers. *IEEE Transactions on Automatic Control*, AC-29(8):696–703, January 1984.

[6] I. Mitrani and P.J.B. King. Multiprocessor systems with preemptive priorities. *Performance Evaluation*, 1(2):118–125, April 1981.

[7] M. Rubinovitch. The slow server problem. *Journal of Applied Probability*, 22(1):205–213, 1985.

[8] M. Rubinovitch. The slow server problem: A queue with stalling. *Journal of Applied Probability*, 22(4):879–892, 1985.

Chapter 10

Networks of queues

All the queues that we have examined in previous chapters were studied in isolation. Jobs arrive at the queue by some 'magic' process, often Poisson, are served, and then disappear for ever from the system. In real life this is rarely the case. The input process to one queue is usually formed by the output of a number of other queues, possibly combined with some entirely new jobs. After service at a queue most jobs join some other queue to await further service. A small proportion may disappear for ever. For example, in a packet switching network, most packets are in transit from one packet switch to another; at each switch, a few new packets may be added representing new messages entering the network, and a few packets may leave the system representing the messages that have reached their destination. This is an appropriate model in many cases, particularly if the population of potential customers is large, and they act independently of one another. However, at other times the total population of potential customers is fixed in advance, and never varies, only the distribution of these customers between queues changes. An example of such a system is a batch computer system with a fixed level of multiprogramming. There are a fixed number of tasks in the system, some will be waiting for the CPU, some will be accessing disks, or waiting for them, and some will be performing other I/O operations. When a task finally terminates, the central scheduler replaces it with another, keeping the total number of active tasks fixed. Other models might require a mixture of these closed and open models. There might be a fixed number of users terminals, each of which can submit a single task to the system, and a separate queue of batch tasks, which are all admitted. The number of terminal tasks is fixed, but the number of batch jobs varies. This chapter will develop a theory of networks of queues that can be used to analyse all of these problems.

10.1 Tandem queues

The simplest possible network, two $M/M/1$ queues in tandem, is shown in Figure 10.1. All jobs arrive at the first server, are served by it, and then go to the second, queueing before either server if necessary. We will assume that there is no limit to the number of jobs that can be waiting for either server. The arrivals at the first queue form a Poisson stream with parameter λ. Service times are exponentially

Figure 10.1 Tandem $M/M/1$ queues

distributed with mean $1/\mu_1$ at server 1 and $1/\mu_2$ at server 2. A job's service time at server 2 is *independent* of its service time at server 1.[†] Intuitively, the condition for the system to reach an equilibrium is that the service rate of the slower server be greater than the arrival rate.

Since all the processes are Markovian, the system state can be represented by a pair of non-negative integers representing the number of customers at each server. The balance equations are simple to derive, given that departures from server 1 go directly to queue 2. Writing $\pi_{i,j} = \Pr(i$ jobs at server 1 and j at server 2),

$$\lambda\pi_{0,0} = \mu_2\pi_{0,1}$$
$$(\lambda + \mu_1)\pi_{1,0} = \lambda\pi_{0,0} + \mu_2\pi_{1,1}$$
$$(\lambda + \mu_2)\pi_{0,1} = \mu_1\pi_{1,0}$$
$$(\lambda + \mu_1 + \mu_2)\pi_{1,1} = \lambda\pi_{0,1} + \mu_1\pi_{2,0} + \mu_2\pi_{1,2}$$
$$(\lambda + \delta(i)\mu_1 + \delta(j)\mu_2)\pi_{i,j} = \lambda\pi_{i-1,j} + \delta(j)\mu_1\pi_{i+1,j-1} + \delta(j+1)\mu_2\pi_{i,j+1}$$

We can try to guess a solution. Since queue 1 is unaffected by what happens at the second queue, the marginal probability of i jobs at the first queue is well known:

$$\Pr(i, \cdot) = \rho_1^i(1 - \rho_1), \quad \rho_1 = \frac{\lambda}{\mu_1}$$

The second queue is more problematic. The arrival process is certainly a renewal process, and assuming that the whole system is stable, then it must have mean rate λ. Its distribution is not known though. If the distribution was Poisson, then the second queue would be independent of the first, and:

$$\Pr(\cdot, j) = \rho_2^j(1 - \rho_2), \quad \rho_2 = \frac{\lambda}{\mu_2}$$

and the joint probability of i jobs in queue 1 and j jobs in queue 2 would be:

$$\pi_{i,j} = \rho_1^i(1 - \rho_1)\rho_2^j(1 - \rho_2)$$

[†]This may seem a rather artificial model, but this independence of service times at different nodes (or even at subsequent visits to the same node) is fundamental to the analysis. The analysis of even this simplest case is extremely complicated in the event that a job has the same service requirement at the two servers.

This certainly satisfies the balance equations so the queues act as if they are independent. The second queue appears to be ignorant of the presence of the first—the state of the second queue is identical to that of an isolated $M/M/1$ queue with input parameter λ. It appears as if the output from an $M/M/1$ queue forms a Poisson stream with the same rate as the input rate, λ. Although the queue length probabilities are the same in the second queue as they would be if the output from the first queue was Poisson, that equality does not *prove* that the output stream is Poisson. We shall now prove that this is the case.

What is the interval between successive departures from an $M/M/1$ queue? If the previous departure left jobs in the queue, then the next departure will occur when the job now starting service finishes its service. In this case, the interdeparture interval is distributed identically to the service time. If the previous departure left the queue empty, the interval will be an interarrival interval plus a service interval. This interval will be distributed as the sum of the two random variables. The Laplace transform of a sum of two random variables is the product of the Laplace transforms of the individual variables. Assuming that the arrival rate is λ, the service rate μ, and $\rho = \lambda/\mu$, we find that the Laplace transform of the interdeparture distribution, $D^*(s)$, is given by:

$$D^*(s) = (1 - \rho)A^*(s)B^*(s) + \rho B^*(s)$$

Now for an exponential distribution with parameter μ, the Laplace transform is:

$$B^*(s) = \frac{\mu}{\mu + s}$$

and hence:

$$\begin{aligned} D^*(s) &= (1 - \rho)\frac{\lambda}{\lambda + s}\frac{\mu}{\mu + s} + \rho\frac{\mu}{\mu + s} \\ &= \frac{\lambda}{\lambda + s} = A^*(s) \end{aligned}$$

so the distribution of interdeparture times is identical to the distribution of interarrival times. The second server can be analysed independently of the first. This property of $M/M/1$ queues was first proved by P.J. Burke, and is often referred to as Burke's theorem. Other queues which have the property of preserving the Poisson nature of the arrival stream in the departure stream include the $M/M/c$ queue, and the $M/G/\infty$ queue.

A network of these types of queues which are connected in tandem can be analysed as if each queue was independent. More complex connections between the queues can be allowed, so long as we maintain the Poisson nature of the flows. The merged stream resulting when several independent Poisson streams are joined is itself a Poisson stream, and each output stream after a Poisson stream is split into several streams is itself a Poisson stream, provided that the splitting is done probabilisticly. These statements can be proved by considering the definition of a Poisson stream of rate λ in terms of the probability of an arrival during a short interval δt.

$$\Pr(1 \text{ arrival in } \delta t) = \lambda \delta t + o(\delta t)$$
$$\Pr(0 \text{ arrivals in } \delta t) = 1 - \lambda \delta t + o(\delta t)$$
$$\Pr(\text{multiple arrivals in } \delta t) = o(\delta t)$$

Consider the output process formed by the merging of two independent Poisson streams with parameters λ_1 and λ_2:

$$\Pr(1 \text{ output in } \delta t) = \Pr(1 \text{ arrival in stream 1 and 0 arrivals in stream 2})$$
$$+ \Pr(0 \text{ arrivals in stream 1 and 1 arrival in stream 2})$$
$$= (\lambda_1 \delta t + o(\delta t))(1 - \lambda_2 \delta t + o(\delta t))$$
$$+ (\lambda_2 \delta t + o(\delta t))(1 - \lambda_2 \delta t + o(\delta t))$$

Performing the multiplication, we see that no terms larger than $o(\delta t)$ remain, except $(\lambda_1 + \lambda_2)\delta t$. Similarly, the probability of more than one arrival is $o(\delta t)$, and of no arrivals is $1 - (\lambda_1 + \lambda_2)\delta t + o(\delta t)$. These are the axioms for a Poisson process with rate $\lambda_1 + \lambda_2$.

A similar analysis can be used to show that a Poisson stream can be split into several separate streams, and so long as the choice of which output stream to add a job to is made at random, the output streams are also Poisson. The restriction to random splitting means that if a Poisson stream is split by sending alternate jobs to alternate streams, the output streams will not be Poisson.

Consider a Poisson stream with rate λ, in which successive arrivals are independently assigned to output stream 1 with probability p, and to output stream 2 with probability $q = 1 - p$. Let $N_i(t)$ be the probability of i arrivals being assigned to stream 1 in an interval of length t.

$$N_1(\delta t) = \Pr(1 \text{ arrival in } \delta t \text{ and that arrival chooses stream 1})$$
$$+ \Pr(\text{multiple arrivals in } \delta t \text{ and only 1 arrival chooses stream 1})$$
$$= \Pr(1 \text{ arrival in } \delta t) \cdot \Pr(\text{arrival chooses stream 1}) + o(\delta t)$$
$$= (\lambda \delta t + o(\delta t)) \cdot p + o(\delta t)$$
$$= p\lambda \delta t + o(\delta t)$$

Similarly, the probability of no arrival being allocated to stream 1 is given by:

$$N_0(\delta t) = \Pr(\text{No arrival in } \delta t)$$
$$+ \Pr(1 \text{ arrival in } \delta t \text{ and that arrival chooses stream 2})$$
$$+ \Pr(\text{Multiple arrivals in } \delta t \text{ and all the arrivals choose stream 2})$$

Figure 10.2 $M/M/1$ system with feedback

$$= (1 - \lambda\delta t) + \lambda\delta t \cdot (1 - p) + o(\delta t)$$
$$= (1 - p\lambda\delta t) + o(\delta t)$$

Clearly, $N_i(\delta t) = o(\delta t)$ for $i \geq 2$, and these probabilities are just the axiomatic definition of a Poisson stream with rate $p\lambda$.

Thus a network of queues in which each queue can have several input streams and in which the output of any queue can be split, can still be analysed as if each queue were independent. The only restriction is that feedback must be impossible; that is, it must *not* be possible for jobs which have left a queue to ever visit the same queue later in their passage through the network. These networks are called feed-forward networks.

10.2   Queues with feedback

Feed-forward networks are applicable to a fairly restricted set of circumstances. When we introduce feedback into the network, so that a job may visit the same queue more than once, the flows in the network need not be Poisson any more. Consider the simple network shown in Figure 10.2. Jobs arrive in a Poisson stream, and receive an exponentially distributed service. After service they leave the system with probability p, or return to the queue with probability $q = 1 - p$. Consideration of the steady state diagram shows us that, as far as queue length is concerned, the system behaves like an $M/M/1$ queue with arrival rate λ and service rate $p\mu$.

Although external arrivals form a Poisson stream, the arrivals at the queue do not. They include not only the external arrivals, which are Poisson and independent of the service process, but also the jobs which have been fed back after service. These fed back arrivals only occur at service completion instants and cannot be independent of service times. The actual distribution of arrivals to the queue is non-exponential. Similarly the intervals between jobs leaving the server, and between jobs being fed back, are not exponentially distributed. Only the stream of jobs departing from the system is Poisson. We do not prove this here.

We could attempt to analyse ever more complex forms of feedback, but instead we move boldly to a general case of M queues which give exponential service.

10.3   Jackson networks

Consider a system consisting of M $M/M/1$ queues. Jobs arrive from outside the system joining queue i at rate λ_i. After service at queue i, which is exponentially distributed with parameter μ_i, the job either leaves the system with probability p_{i0}, or goes to queue j, with probability p_{ij}. Clearly, $\sum_{j=0}^{M} p_{ij} = 1$, since each job must go somewhere.

The system as a whole is clearly Markovian, with the state being given by the vector of non-negative integers representing the queue lengths at each node. The vector $\mathbf{n} = (n_1, n_2, \ldots, n_M)$ represents the state of having n_1 jobs in queue 1, n_2 jobs in queue 2, etc. We assume that a steady state is reached and write $\pi(\mathbf{n})$ for the steady state probability of state $\mathbf{n}$. The balance equations can be easily written down:

$$\pi(\mathbf{n})(\sum_{i=1}^{M} \lambda_i + \sum_{i=1}^{M} \delta(n_i)\mu_i) = \sum_{i=1}^{M} \delta(n_i)\lambda_i\pi(\mathbf{n} - \mathbf{e}_i)$$

$$+ \sum_{i=1}^{M} p_{i0}\mu_i\pi(\mathbf{n} + \mathbf{e}_i) \qquad (10.1)$$

$$+ \sum_{i=1}^{M}\sum_{j=1}^{M} \delta(n_j)p_{ij}\mu_i\pi(\mathbf{n} + \mathbf{e}_i - \mathbf{e}_j)$$

where $\delta(k)$ is the Kronecker delta function, equal to 1 if $k > 0$ is satisfied, and 0 otherwise, and $\mathbf{e}_j$ is the vector with all components zero, except the jth, which is 1. Notice that the term $\delta(n_i)p_{ii}\mu_i\pi(\mathbf{n})$ appears on both sides of the equation, representing jobs which leave node i and immediately rejoin node i. First we find the total throughput of node i. It consists of those jobs which arrive at node i from outside the network, at rate λ_i, plus all those jobs which are transferred to node i after completing service at node j for all nodes j in the network. If γ_j is the throughput of node j, then the rate at which jobs arrive at node i from node j is $p_{ji}\gamma_j$. Hence the throughput of node i, γ_i must satisfy the traffic equations:

$$\gamma_i = \lambda_i + \sum_{j=1}^{M} p_{ji}\gamma_j \qquad (10.2)$$

These equations have a solution provided that p_{i0} is non-zero for some i. Defining ρ_i as γ_i/μ_i, the state probabilities are given by:

$$\pi(\mathbf{n}) = \prod_{i=1}^{M} \rho_i^{n_i}(1 - \rho_i)$$

This can be easily verified by substitution in the balance equations (10.1). The queues can be analysed as if they are independent $M/M/1$ queues. It is important to note that the steady state probabilities behave *as if* the flows were Poisson, but the flows are *not*. The network state probability is the product of the probabilities of the individual queues. These *product form* networks are often known as 'Jackson networks' after J.R. Jackson, who first proved this property. They are open networks, that is, the population of jobs is not fixed, but can vary as new jobs arrive in the system and as jobs leave the system.

10.4   Closed queueing networks

Closed networks, in which a fixed number of jobs move between queues have also been studied. The machine repairman model is a simple example of a closed queueing network. The n machines are alternately operative, for a period which is exponentially distributed, and then broken down, when they are repaired by a single repairman who takes an exponentially distributed time to effect each repair, and works on one machine at a time. The operational state machines can be thought of as being at an $M/M/\infty$ queue, and the broken ones as at an $M/M/1$ queue.

An early application of closed queueing networks was the 'central server model', which is intended to model batch systems with a fixed level of multiprogramming. Each job represents one program in the multiprogramming load, and after each visit to the CPU a job visits one of the I/O devices before returning to the CPU. When a job eventually leaves the system, it is replaced by another, keeping the total number constant. This situation can be modelled by the network shown in Figure 10.3. Queue 0 represents the CPU, and jobs leaving it go to I/O device i with probability p_i. With probability $q = 1 - \sum_{i=1}^{M} p_i$ the job leaves the system, and a new job joins the CPU queue; this is the so called 'new job loop'. The service times at the CPU and the I/O devices are exponentially distributed. Again the state of the system is given by the number of jobs at each server, and the system is clearly Markovian. Unlike the open case though, the total number of jobs in the system must remain constant. The balance equations are similar to those for the open case except that all the terms involving jobs entering or leaving the system are zero.

$$\pi(\mathbf{n})\left((1-q)\delta(n_0)\mu_0 + \sum_{j=1}^{M} \delta(n_j)\mu_j\right) =$$

$$\sum_{j=1}^{M} [\delta(n_j)\mu_0 p_j \pi(\mathbf{n} + \mathbf{e}_0 - \mathbf{e}_j) + \delta(n_0)\mu_j \pi(\mathbf{n} - \mathbf{e}_0 + \mathbf{e}_j)] \qquad (10.3)$$

The balance equations form a finite set, and can be solved numerically in a number of ways. We shall examine some numerical solution techniques in chapter 13. However

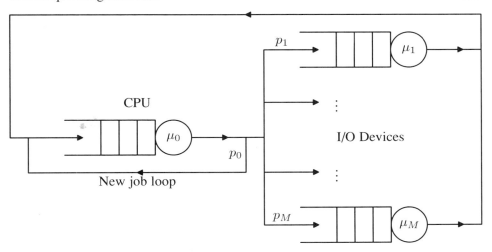

Figure 10.3 Central server system

we try a general solution of the form $\pi(\mathbf{n}) = G \prod_{i=0}^{M} \rho_i{}^{n_i}$, where G is an arbitrary constant, $p_0 = 1$, and $\rho_i = p_i / \mu_i$. This satisfies the balance equations, so we must choose the arbitrary constant G so that the probabilities sum to 1. G is known as the *normalisation constant*.

   This formula seems to offer an easy way to calculate the state probabilities, for we could simply sum $\prod_{i=0}^{M} \rho_i{}^{n_i}$ over all the possible assignments of N jobs to $M + 1$ queues and hence calculate G from the normalisation condition. Notice that the values of ρ_i can all be multiplied by a scaling factor without affecting the validity of the solution to the balance equations. The only effect is to change the value that the normalisation constant, G, must have. Although this approach can be followed, there are severe drawbacks. A system of N jobs and $M + 1$ queues has $\binom{N+M}{M}$ states. (To see this, represent the system state by $N + M$ bits, N zero bits, and M one bits. A zero bit represents a job in the queue i when there are i one bits to the left of the bit in question. The number of possible states is the number of ways of choosing M one bits among $N + M$ bits. This number, although finite, rapidly increases with N and M. With 5 queues and 10 jobs, there are $\frac{14!}{10!4!} = 1001$ states. As the number of jobs increases, the number of states increases very rapidly. Although it might be feasible to carry out the state probability calculations for all states, the accumulated round-off error in G would probably make the results meaningless. It is also worth observing that it is not usually the individual state probabilities that we are interested in, but marginal probabilities of the queue lengths at particular nodes, or higher level measures such as utilisations of servers, and mean queue lengths. In the next chapter we shall return to the problem of calculating G and the other measures, but for now we shall consider generalisations of the network model.

10.5   More general networks

In the networks we have seen so far, all jobs are statistically identical, and all the nodes offering service only provide an exponentially distributed service. In real life, neither of these properties is likely to hold. For example, in a computer system there will be jobs which are CPU bound, and only use a very small amount of I/O, as well as jobs which are I/O bound and only process for short times between successive I/O requests. The desire to incorporate this diversity led to the analysis of networks with different classes of job. Each class of job may have its own routing through the network, and service demands. Different distributions of service requirement were also studied. This work culminated in the so-called BCMP theorem, named after the initials of the authors of the paper which unified the various results. We shall discuss the theorem at some length, and then give some later extensions.

10.6   Coxian distributions

The great simplicity of the exponential distribution comes from its memoryless property; that is, we do not need to know the time since a job started service in order to know the distribution of the time until it will end service. Erlang realised the great simplicity of analysis that this allowed, and introduced the form of distribution now known by his name. In an Erlang-k distribution, there are k identical exponential servers, and the job in service visits them one after another. The state of the combined server is now given by the stage of service that a job has reached. Only one job will be in service at any time. The introduction of the Erlang-k distribution allows a wider range of service distributions to be studied, but all Erlang-k distributions have coefficient of variance less than 1. Even if the stages are allowed to have different rates of service, the variance will not exceed 1. In order to consider services which have higher variance, other approximations must be used.

The hyper-exponential distribution is another distribution which can be represented exactly by a set of exponential stages. In this case the stages are in parallel, and only one of them is chosen at random. For example, an H_2 distribution might consist of an exponential distribution with parameter a with probability p and an exponentially distributed time with parameter b with probability $1 - p$. Hyper-exponential distributions can have variances which are larger than 1, but never less than 1. It is a straightforward exercise to analyse queues using either the Erlang-k or the hyper-exponential distributions of service times for jobs.

Figure 10.4 Construction of a Coxian server

Coxian distributions are constructed similarly to Erlang distributions. A series of exponential servers are visited one after another. The ith stage of service has parameter μ_i. After receiving service at stage i, a job will continue to stage $i+1$ with probability b_i, and leave the server completely with probability $1 - b_i$. If there are k stages, then $b_k = 0$. The full generality of a Coxian distribution includes a probability $1 - b_0$ of leaving without any service at all. Figure 10.4 shows a 3 stage Coxian server. At most one job can be in any of the stages at one time. Writing A_j for $(1 - b_j) \cdot \prod_{i=0}^{j-1} b_i$, the mean of a Coxian distribution with k stages is given by:

$$\frac{1}{\mu} = \sum_{i=1}^{k} A_i \sum_{j=1}^{i} \frac{1}{\mu_j} = \sum_{i=1}^{k} \frac{A_i}{1 - b_i} \frac{1}{\mu_i}$$

Cox was able to show that *any* distribution with a rational Laplace transform can be approximated as closely as desired using this construction. Allowing the job to skip stages or to loop back and visit stages a second time adds no generality to the forms of distribution that can be approximated. This enables a network of servers which give generally distributed service to be transformed to a network of exponential servers (albeit with restrictions on the number of jobs in sub-networks) with identical characteristics.

10.7   Service disciplines

Let us now consider service disciplines other than FCFS. Last come first served, preemptive resume, otherwise known as LCFSPR, is a discipline in which a job arriving at the server displaces any job which is already receiving service. It executes until either it completes its service or another job arrives in which case it is displaced. When a job completes service and leaves the server, the job which it displaced resumes execution at the point at which it was interrupted. Consider an $M/G/1$ system in which jobs are of R classes, each of which has its own Coxian service time requirement. Jobs of class r arrive from outside the system at rate λ_r and require service which has a Coxian distribution with L_r stages, stage s being exponential with parameter $\mu_{(r,s)}$, and probability of continuing to stage $s + 1$ after stage s being $b_{(r,s)}$. Defining $A_{(r,s)}$ as above, the mean service requirement of class r jobs is given by:

$$1/\mu_r = \sum_{s=1}^{L_r} \frac{A_{(r,s)}}{1 - b_{(r,s)}} \frac{1}{\mu_{(r,s)}}$$

We now consider a single server queue with a LCFSPR discipline. The state of the queue can be given by a vector of pairs of integers, one pair for each job in the system, in LCFS order. The first element of each pair represents the class of that job, and the second element stands for the stage of service reached by that job. $((c_1, s_1), (c_2, s_2), \ldots, (c_n, s_n))$ represents the system when there are n jobs in the queue. Job n, which is of class c_n, is the most recently arrived, and is currently receiving service from stage s_n. With this state description, the system is Markovian, and we can write down the balance equations. The system will leave that state at rate $\mu_{(c_n, s_n)} + \sum_{r=1}^{R} \lambda_r$, representing the total arrival rate of jobs of all classes and the rate of completing service of the current stage. Entry to this state occurs from states with one more customer, $(\ldots, (c_{n+1}, s_{n+1}))$ at rate $(1 - b_{(c_{n+1}, s_{n+1})})\mu_{(c_{n+1}, s_{n+1})}$, representing the job finishing service and leaving; from states with the same number of customers at rate $b_{(c_n, s_n-1)}\mu_{(c_n, s_n-1)}$ representing the previous stage of the same job finishing; and, if $s_n = 1$ at rate $\lambda_{c_n} b_{(c_n, 0)}$ from the state with only $n - 1$ jobs in the system. Although the state description is rather unwieldy, it is easy to write the balance equations. By inspection, we try the solution:

$$\Pr((c_1, s_1), (c_2, s_2), \ldots, (c_n, s_n)) = (1 - \rho) \prod_{i=1}^{n} \lambda_{c_i} \cdot \frac{A_{(c_i, s_i)}}{(1 - b_{(c_i, s_i)})\mu_{(c_i, s_i)}} \quad (10.4)$$

where $\rho = \sum_{r=1}^{R} \rho_r$, and $\rho_r = \lambda_r/\mu_r$. This state description is actually rather too detailed for most purposes; usually we are not interested in the exact order of the jobs in the queue, nor will we be interested in the exact stage which each job has reached. First, let us find the probability of the queue consisting of jobs of classes $c_1, c_2, \ldots, c_n$ in that order, but without distinguishing the stage of service that each job has reached. Summing over all relevant assignments of stages in equations (10.4), we find:

$$\Pr(c_1, c_2, \ldots, c_n) = (1 - \rho) \prod_{i=1}^{n} \rho_{c_i}$$

The state description can be simplified further, to give the probability that there are k_1 jobs of class 1, k_2 jobs of class 2, and so on. In this case,

$$\Pr(k_1, k_2, \ldots, k_R) = (1 - \rho)k! \prod_{r=1}^{R} \frac{\rho_r^{k_r}}{k_r!} \quad (10.5)$$

where $k = k_1 + k_2 + \cdots + k_R$. If the length of queue, independent of the exact classes of the jobs involved, is adequate for our purposes, then the probability simplifies still further to give:

$$\Pr(k) = (1 - \rho)\rho^k$$

So despite all the extra complexity of the service discipline, the queue length probability is exactly the same as it would be in the simple $M/M/1$ system with a single class of jobs with traffic intensity ρ. Although the service distributions of the individual classes can be arbitrarily complex, and different for each class, the probability of the queue having a particular number of jobs of particular classes is totally independent of the form of the distribution, and depends only on the mean service time requirements.

We now ask what other service disciplines have this property.

10.7.1   Processor sharing discipline

The discipline known as *processor sharing* is used as an approximation to a time sharing system in which each job receives a small quantum of service and is then suspended until every other job has received an identical quantum of service in a round-robin fashion. As the quantum becomes smaller and smaller, in the limit, each job is receiving service at $1/n$ of the server's capacity when there are n jobs in the queue.

As with the LCFSPR queue, we can allow jobs of different classes to have different service time distributions, so long as they are Coxian distributed. Since all jobs are receiving service simultaneously, we do not need to record the order of arrival. The detailed state can be described by the vector of vectors $(\mathbf{k}_1, \mathbf{k}_2, \ldots, \mathbf{k}_R)$ where $\mathbf{k}_r = (k_{(r,1)}, k_{(r,2)}, \ldots, k_{(r,L_r)})$ is the vector of the jobs of class r. $k_{r,s}$ represents the number of jobs of class r which are at stage s. As before, we write k_r for the total number of jobs of class r ($\sum_{s=1}^{L_r} k_{(r,s)}$), and k for the total number of jobs of all classes. Arrivals in this state occur at rate $\lambda_r b_{(r,0)}$ and cause a transition to the state with one more job at stage 1 of class r. A job of class r at stage s finishes that stage at rate $\mu_{(r,s)}/k$, and with probability $b_{(r,s)}$ it continues to stage $s + 1$, and otherwise it leaves. The following solution satisfies the balance equations.

$$\Pr(\mathbf{k}_1, \mathbf{k}_2, \ldots, \mathbf{k}_R) = (1 - \rho)k! \prod_{r=1}^{R} \left[\prod_{s=1}^{L_r} (\lambda_r A_{rs}/((1 - b_{rs})\mu_{rs})^{k_{rs}} k_{rs}!) \right]$$

Aggregating the states that have the same number of jobs of each class, we find:

$$\Pr(k_1, k_2, \ldots, k_R) = (1 - \rho)k! \prod_{r=1}^{R} \rho_r^{k_r}/k_r!$$

which is identical to the expression for LCFSPR (10.5).

10.7.2   Server per job discipline

This discipline can be used to model situations where each job proceeds independently of the other jobs in the system, receiving a random service. The $M/M/\infty$ queue has already been studied in section 9.2. We are now going to analyse the $M/G/\infty$ queue.

As with the other disciplines, we allow jobs of different classes to have different service time distributions, so long as they are Coxian distributed. The detailed state can be described in exactly the same way as for processor sharing; all jobs are receiving service simultaneously. Transitions caused by arrivals from outside the system occur at the same rate as before; transitions to other states occur at k times the rate they did in the PS case. The steady state probability is given by:

$$\Pr(\mathbf{k}_1, \mathbf{k}_2, \ldots, \mathbf{k}_R) = \mathrm{e}^{-\rho} \prod_{r=1}^{R} \prod_{s=1}^{L_r} \left[(\lambda_r A_{rs}/((1-b_{rs})\mu_{rs})^{k_{rs}} k_{rs}!) \right]$$

Aggregating the states that have the same number of jobs of each class, we find:

$$\Pr(k_1, k_2, \ldots, k_R) = \mathrm{e}^{-\rho} \prod_{r=1}^{R} \rho_r^{k_r}/k_r!$$

Again states can be aggregated still further to give the probability of k jobs being present as:

$$\Pr(k) = \mathrm{e}^{-\rho} \frac{\rho^k}{k!}$$

We now have three different scheduling disciplines that have the property that the queue length distributions are *insensitive* to the form of the service time distribution, and depend only on the mean service time. What is so special about these disciplines and are there any others that have this property?

10.8   Local balance

The three disciplines which were discussed above have the property known as *local balance*. The global balance equations corresponding to the flows out of and into one state consist of many terms. These terms can be collected so that all terms relating to state changes caused by flow of jobs into the state caused by arrivals of jobs of class i at stage j are together, and all terms relating to changes of state caused by the flow of jobs of class i out of stage j are also together. In those disciplines which have local balance, the terms due to flow into the state caused by jobs of class i entering stage j exactly cancel those terms due to the flow out of the state caused by jobs of class i leaving stage j. The local balance equations are formed by equating these two sets of terms. Any solution to the local balance equations must be a solution of the global balance equations, since they are just the sum of the local balance equations over all classes and stages. A solution to the global balance equations need not satisfy the local balance equations.

The global balance equations for the central server network, (10.3), can be decomposed in this way. In state $\mathbf{n}$, the term corresponding to leaving the state because of a departure from node j to the central server is just $\pi(\mathbf{n})\mu_j$, for $j \neq 0$. Departures from the central server occur at rate $\pi(\mathbf{n})(1 - q)\mu_0$. Entry to the state $\mathbf{n}$ caused by arrivals at node j occurs when a job leaves the central server, an event which occurs at rate $p_j\mu_0$ from the state $\mathbf{n} + \mathbf{e}_0 - \mathbf{e}_j$. Entry to the state caused by arrivals at the central server occurs at rate μ_j from state $\mathbf{n} + \mathbf{e}_j - \mathbf{e}_0$. Equating these terms gives the local balance equations:

$$\pi(\mathbf{n})(\delta(n_j)\mu_j) = \delta(n_j)p_j\mu_0\pi(\mathbf{n} + \mathbf{e}_0 - \mathbf{e}_j) \quad j \neq 0$$

$$\pi(\mathbf{n})(\delta(n_0)\mu_0) = \sum_{j=1}^{M} \delta(n_0)\pi(\mathbf{n} + \mathbf{e}_j - \mathbf{e}_0)\mu_j$$

The global balance equations are just the sum of these local balance equations. It is readily confirmed that the general product form $\pi(\mathbf{n}) = G \prod_{i=0}^{M} \rho_i^{n_i}$ satisfies the local balance equations. Taking the equation for node j and state $\mathbf{n}$, we have, after dividing both sides by common factors,

$$\rho_j\mu_j = p_j\mu_0\rho_0$$

Since $\rho_j = p_j/\mu_j$ and $p_0 = 1$, this is an identity. Similarly for the equation relating to the central server,

$$(1 - q)\rho_0\mu_0 = \sum_{j=1}^{M} \rho_j\mu_j$$

Since $q = 1 - \sum p_j$, this reduces to an identity also.

10.9 BCMP theorem

We are now in a position to state the fundamental theorem of queueing networks whose steady state distributions satisfy a product form; the BCMP theorem. The theorem unifies all of the results previously presented in this chapter. In order to reduce confusion, we shall have to change our nomenclature slightly. Networks consist of a number of *nodes* which each have a number of queues and servers. The networks covered by the theorem allow multiple classes of job, each with its own routing probabilities between the various nodes in the network. Each class can be open, that is allow arrivals and departures, or closed, that is have a fixed number of jobs, independently of the other classes. Jobs can change class when they move from one node to another.

The nodes in the network can be of four types;

Type 1: Jobs form a single queue which is served in FCFS order. All classes of job must have the same exponential service distribution. The rate of service can depend on the length of the queue. This feature is usually used to model multiple server queues, by making the rate increase linearly with queue length, until the maximum rate is reached.

Type 2: A processor sharing node. All jobs join a single queue, and are served simultaneously, sharing the server equally. Jobs of different classes may have different service time distributions, which must be Coxian. The speed of the server may depend on the length of the queue, in the same way as for type 1 nodes.

Type 3: Server per job node. Each job at this type of node receives its service without waiting. Jobs of different classes may have different service time distributions, which must be Coxian. As with type 1 and type 2 nodes, the rate of service can also vary with queue length.

Type 4: LCFSPR node. Newly arrived jobs displace the job currently in service, which will resume at the point of interruption when all jobs which have arrived after it have been served. Jobs of different classes may have different service time distributions, which must be Coxian. The rate of service can vary with queue length.

We assume that there are M nodes. When a job of class r has been served at node i it is transferred to node j and class s with probability $p_{ir;js}$; with probability $p_{ir;0}$ the job leaves the network. Depending on which transfers between classes are possible, the set of node-class pairs can be partitioned into a number of non-intersecting subsets. These subsets are often called *routing chains*. Two node-class pairs will be in the same routing chain if and only if there is a positive probability that a job which is in one of the pairs will be in the other pair before it leaves the network, or has been in the other pair before it reached its current pair.

Arrivals can occur in one of two ways, but not both. The first possibility is a single Poisson arrival stream, at rate $\lambda(k)$, which may depend on the total number of jobs in the network. Jobs arriving in this way are assigned to node i with class r with probability $p_{0,ir}$. The alternative is for each routing chain to have its own Poisson arrival stream whose rate may depend on the number of jobs in that routing chain.

Whichever form of external arrivals are assumed, in order to find the steady state distributions, the following equations must be solved for each routing chain C:

$$\sum_{i,r \in C} v_{ir} p_{ir;js} + p_{0;js} = v_{js} \quad j,s \in C \tag{10.6}$$

If there are no external arrivals in chain C, the equations are singular, but the values of v_{ir} for that chain can be determined up to a multiplicative constant. The values of v_{ir} represent the relative frequency of visits to that node-class pair by jobs in the (closed) routing chain C, and are known as the *visit ratios*. When there are external arrivals to some node of a routing chain, the equations (10.6) have a unique solution, determining v_{ir} as the expected number of visits by a class r job to node i.

As with the single queue open disciplines that we studied in the previous sections, it is possible to define very detailed system states to make the system Markovian, find the steady state distribution, and then aggregate to find the steady state of the (non-Markovian) aggregated states. We shall omit that stage, which is identical to the derivation above, merely remarking that the local balance properties are used to simplify the verification that the global balance equations are satisfied. Moving straight to a state description where queue i is described by a vector, $\mathbf{k}_i = (k_{i1}, k_{i2}, \ldots, k_{iR})$, with k_{ij} representing the number of jobs in the class j at queue i. As before, we use k_i to denote the total number of jobs at queue i, $k_i = \sum_{r=1}^{R} k_{ir}$. k represents the total population of the network, $k = \sum k_i$. The steady state probabilities then satisfy:

$$\Pr(S = (\mathbf{k}_1, \mathbf{k}_2, \ldots, \mathbf{k}_M)) = \frac{1}{G} d(S) f_1(\mathbf{k}_1) f_2(\mathbf{k}_2) \ldots f_M(\mathbf{k}_M) \qquad (10.7)$$

where G is an arbitrary constant, $d(S)$ is a function of the number of customers in the system, and $f_i(\mathbf{k}_i)$ is a function which depends not only on the customers present at node i, but also on what type of node it is.

When node i is type 1 (FCFS), 2 (PS), or 4 (LCFSPR), then $f_i(\mathbf{k}_i) = k_i! \prod_{r=1}^{R} (1/k_{ir}!)[v_{ir}/\mu_{ir}]^{k_{ir}}$.

When node i is type 3 (IS), then $f_i(\mathbf{k}_i) = \prod_{r=1}^{R} (1/k_{ir}!)[v_{ir}/\mu_{ir}]^{k_{ir}}$.

Those two expressions are appropriate formulas if the service rates are independent of the number of jobs in the queue. If the service rate when there are j jobs is $s(j)\mu_{ir}$, then the term $\prod_{r=1}^{R} [1/\mu_{ir}]^{k_{ir}}$ should be replaced by $\prod_{j=1}^{k_i} 1/(s(j)\mu_{ir})$.

The term $d(S)$ is governed by the external arrivals to the system. If all routing chains are closed, it will have the value 1. Otherwise, if arrivals depend on the total population in the network, it has the value $\prod_{j=0}^{k-1} \lambda(j)$ and if arrivals depend on the populations of individual routing chains then it is equal to $\prod_{C=1}^{N_C} \prod_{j=0}^{k_C-1} \lambda_m(j)$ where k_C is the number of jobs in the routing chain C, and N_C is the number of routing chains.

The arbitrary constant G is chosen so that the probabilities (10.7) sum to 1; it is known as the normalisation constant. As we observed above when discussing the central server model in section 10.4, the naive method of calculating G will quickly become infeasible because of the large number of states to be considered, even if the numerical problems of summing such a large number of small quantities accurately could be solved. In the next chapter we consider a number of methods for calculating either G, and from it the various measures of performance such as queue lengths, or for calculating these measures directly. First we outline a number of extensions to the applicability of the BCMP theorem.

10.9.1   Extensions to BCMP theorem

Although the BCMP theorem was a major breakthrough, and covers a large number of cases, there has been a large amount of subsequent research with the aim of finding product form solutions for less restricted networks.

Under certain conditions, product form will still hold for the solution of a network with population constraints. A network which is open to the arrival of new jobs can be constrained to a maximum number of jobs present in the net. This can easily be accommodated in the BCMP framework using the ability to have state dependent arrival rates. Lower limits on the network population can also be included in a network and the solution may still have product form. The departure of a job can *trigger* a new arrival if the population would otherwise drop below the constraints. A sufficient condition for the network to retain its product form solution is that if the state dependent arrival rate of class c customers to the network is 0 when the population is $(n_1, n_2, \ldots, n_c, \ldots, n_R)$, then a departure of a class c job from state $(n_1, n_2, \ldots, n_c + 1, \ldots, n_R)$ should trigger an arrival of a class c job to the network. That is all feasible changes in the network population vector must be reversible.

State dependent routing is another extension that has been considered by a number of authors. The restrictions needed to maintain product form solution mean that state dependent routing applies only where a choice is being made between parallel sub-networks. Parallel sub-networks are mutually disjoint subsets of the nodes which can only be entered from one node, and when jobs leave parallel sub-networks they all go to the same node.

In addition to the original four service disciplines, a number of other scheduling schemes have been shown to preserve the product form of the solution. Random order of service, so long as all classes of jobs have the same exponential distribution of service times is one. This is perhaps unsurprising, since if all jobs have the same exponential service time distribution, then they will be statistically indistinguishable, and it will not matter in the long run which job is chosen for service.

Load balancing nodes can be introduced to a network, and still maintain a product form solution. They can have several servers, and service a number of queues. When a server becomes idle, it will choose to serve one of the queues associated with the node with a probability that depends on all the queue lengths. If all queues are of equal length, then the queue to be served will be chosen at random. If the queues are of unequal length, then the probabilities are such as to make a long queue more likely to be served than a short one. The intention is to attempt to keep all the queues at the node approximately equal in length.

10.10   Exercises

1. Derive the local balance equations for the Jackson network. Show that after substitution of the product form solution, they reduce to the traffic equations (10.2).

10.11   Further reading

Initial studies of queues all concerned isolated systems. Tandem queues were solved in the form shown here by Burke[3]. A closed system of Markovian queues in which the output of queue i became the input of queue $i+1$, a cyclic system, was studied by Koenigsberg [11, 12]. R.R.P. Jackson was responsible for a comprehensive analysis of queues in tandem[10].

J.R. Jackson first analysed networks of $M/M/1$ queues[8]. In a later paper, he extended the work to allow state dependent arrival rates and multiple servers[9]. Closed queueing networks are usually attributed to Gordon and Newell[6, 7], although Koenigsberg's result is a special case of this, and Gordon and Newell's result is actually a special case of Jackson's general result[9]. Strangely enough, Posner and Bernholz[14, 15] published related results almost simultaneously.

The original BCMP result was announced in a paper by Baskett, Chandy, Muntz, and Palacios[2]. It unified a number of results that each author had derived independently. Numerous authors have added to the understanding of the conditions under which product form solutions to queueing networks can be expected. Towsley[17] showed conditions under which state-dependent routing could be allowed, while retaining a product form solution. Lam[13] analyses networks with population constraints, neither open or closed, and shows them to be product form under suitable conditions. Random order of service was shown to be admissible by Spirn[16]. Afshari, Bruell, and Kain[1] demonstrated that load balancing nodes could also be allowed.

The concept of *station balance* was introduced by Chandy to give sufficient conditions for a product form[4, 5].

10.12   Bibliography

[1]   P.V. Afshari, S.C. Bruell, and R.Y. Kain. On the load balancing bus accessing scheme. *IEEE Transactions on Computers*, C-32(8):766–770, January 1983.

[2]   F. Baskett, K.M. Chandy, R.R. Muntz, and F. Palacios-Gomez.   Open, closed and mixed networks of queues with different classes of customers. *Journal of the ACM*, 22(2):248–260, April 1975.

[3]   P.J. Burke.   The output of a queueing system. *Operations Research*, 4(6):699–704, November/December 1956.

[4]   K.M. Chandy, J.H. Howard, and D.F. Towsley. Product form and local balance in queueing networks. *Journal of the ACM*, 24(2):250–263, April 1977.

[5]   K.M. Chandy and A.J. Martin. A characterization of product-form queueing networks. *Journal of the ACM*, 30(2):286–299, April 1983.

[6]   W.J. Gordon and G.F. Newell. Closed queueing networks with exponential servers. *Operations Research*, 15(2):254–265, March/April 1967.

[7]   W.J. Gordon and G.F. Newell. Cyclic queueing systems with restricted length queues. *Operations Research*, 15(2):266–277, March/April 1967.

[8]   J.R. Jackson. Networks of waiting lines. *Operations Research*, 5(4):518–521, July/August 1957.

[9]   J.R. Jackson. Jobshop-like queueing systems. *Management Science*, 10(1):131–142, 1963.

[10]   R.R.P. Jackson. Queueing systems with phase-type service. *Operational Research Quarterly*, 5(4):109–121, 1954.

[11]   E. Koenigsberg. Cyclic queues. *Operational Research Quarterly*, 9(1):22–35, 1958.

[12]   E. Koenigsberg. Finite queues and cyclic queues. *Operations Research*, 8(2):246, March/April 1960.

[13]   S.S. Lam. Queueing networks with population size constraints. *IBM Journal of Research and Development*, 21(4):370–378, July 1977.

[14]   M.J.M. Posner and B. Bernholz. Closed finite queueing networks with time lags. *Operations Research*, 16(5):962–976, September/October 1968.

[15]   M.J.M. Posner and B. Bernholz. Closed finite queueing networks with time lags and with several classes of unit. *Operations Research*, 16(5):977–985, September/October 1968.

[16]   J.R. Spirn. Queueing networks with random selection for service. *IEEE Transactions on Software Engineering*, SE-5(3):287–289, 1979.

[17]   D.F. Towsley. Queueing network models with state-dependent routing. *Journal of the ACM*, 27(2):323–337, April 1980.

Chapter 11

Computational algorithms for product form queueing networks

We have seen that BCMP networks have a simple steady state probability solution. When the network is open, and all classes of jobs have populations which are unbounded, the problem of calculating the steady state probabilities is primarily algebraic, and closed forms are easily found. If at least some of its routing chains are closed, with a fixed or bounded population of jobs, then the calculation of the steady state probabilities depends on calculating the normalisation constant G.

When G is known, then the probabilities can be calculated, and from them the performance measures such as mean queue lengths, utilisations, and throughputs of the various nodes in the network. We shall examine several methods for calculating the performance measures, either by way of G, or directly, and comment on their numerical properties.

Single class closed queueing networks will be considered first, and then extensions to the algorithms to accommodate multiple class closed networks will be outlined. Finally, extension to mixed networks with some classes open, and some closed, will be outlined.

11.1 Convolution algorithm

We start by examining a single class closed queueing network, with M nodes and a population of N jobs. All the nodes are assumed to be fixed rate servers of type 1. μ_i is the parameter of the exponential service given at node i. We shall write ρ_i for v_i/μ_i, where v_i is the visit ratio of node i. It is found by solving the routing equations (10.6). Since there are no external arrivals, the equations are singular, but if one of the v_i, v_k say, is arbitrarily given some non-zero value then all the other v_j are determined. Notice that this means that the value of ρ_j is not uniquely determined, but depends on the arbitrary value assigned to v_j. ρ_i is sometimes called the *relative utilisation* of node i.

Since all processes involved in state change are Markovian, the state of the system can be expressed as an M-vector of integers, $(n_1, n_2, \ldots, n_M)$, with n_i representing the number of jobs at node i. The BCMP theorem then gives the steady state probabilities for a network of fixed rate servers as:

$$\Pr(n_1, n_2, \ldots, n_M) = \frac{1}{G} \prod_{i=1}^{M} \rho_i^{n_i}, \quad n_1 + n_2 + \cdots + n_M = N$$

G is the normalisation constant, chosen to make the probabilities sum to 1. It is the sum over all feasible assignments of jobs to nodes of the product term $\prod_{i=1}^{M} \rho_i^{n_i}$. That is:

$$G = \sum_{\mathbf{n} \in \mathcal{S}} \prod_{i=1}^{M} \rho_i^{n_i}$$

where $\mathbf{n} = (n_1, n_2, \ldots, n_M)$ and $\mathcal{S} = \{\mathbf{n} | n_1 + n_2 + \cdots + n_M = N\}$. It was remarked in the previous chapter that the number of feasible states is $\binom{N+M-1}{M-1}$, and this number is such that for any reasonable values of M and N, direct summation is impractical.

Consider the (improper) generating functions:

$$\gamma_i(z) \stackrel{\triangle}{=} \sum_{j=0}^{\infty} f_i(j) z^j$$

$$= 1 + \rho_i z + \rho_i^2 z^2 + \rho_i^3 z^3 + \cdots$$

$$= \frac{1}{1 - \rho_i z} \tag{11.1}$$

and also the (improper) generating function $\mathcal{G}_j(z)$:

$$\mathcal{G}_j(z) = \prod_{i=1}^{j} \gamma_i(z)$$

The generating functions are improper because they do not (necessarily) correspond to probabilities, nor do they have value 1 at 1. The coefficient of z^k in $\mathcal{G}_2(z)$ is $\sum_{i=0}^{k} \rho_1^i \rho_2^{k-i}$, which is the sum of all possible assignments of k jobs between two processors. Similarly, the coefficients of $\mathcal{G}_3(z)$ correspond to the sum of products with jobs circulating amongst three processors. Hence the coefficient of z^N in $\mathcal{G}_M(z)$ is just the value of the normalisation constant G when there are N jobs.

We want a method of calculating the coefficients of $\mathcal{G}_M(z)$. There is an obvious recurrence for calculating $\mathcal{G}_j(z)$:

$$\mathcal{G}_j(z) = \gamma_j(z) \cdot \mathcal{G}_{j-1}(z) \quad j = 2, \ldots, M \tag{11.2}$$

Although the $\gamma_j(z)$ have an infinite set of coefficients, since we are only concerned with the coefficient of z^N in the end, we can treat each as a polynomial of order N. We then have a series of $M - 1$ polynomial multiplications. The coefficients can be found by taking the discrete convolutions of the coefficient sequences of the polynomials,

$$G_j(n) = \text{Coefficient of } z^n \text{ in } \mathcal{G}_j(z) = \sum_{k=0}^{n} f_j(k)G_{j-1}(n-k) \qquad (11.3)$$

Although this is the manner in which the calculation must be done for arbitrary service rate dependencies, in the case currently being considered, with state independent service rates, a more efficient organisation of the calculation is possible. Using the definition of $\gamma_i(z)$ in equation (11.1), we can express the recurrence (11.2) as:

$$\mathcal{G}_j(z) = \frac{1}{1 - \rho_j z} \cdot \mathcal{G}_{j-1}(z),$$
$$\mathcal{G}_j(z) = \mathcal{G}_{j-1}(z) + \rho_j z \mathcal{G}_j(z)$$

In terms of the coefficients, we can express the last recurrence as:

$$G_j(n) = G_{j-1}(n) + \rho_j \cdot G_j(n-1)$$

Initialising $G_0(i) = 0, i > 0$ and $G_0(0) = 1$, the required coefficient $G_M(N)$ can be calculated in $O(MN)$ operations, if all the nodes in the networks serve at fixed rate. This is known as the *convolution algorithm* or *Buzen's algorithm*. A minor practical improvement in the algorithm can be made by starting the initialisation with node 1, so that $G_1(n) = \rho_1 \cdot G_1(n-1)$, for $n > 0$ and $G_1(0) = 1$. In the remainder of this chapter, G with no subscript or argument is to be understood as standing for $G_M(N)$, and $G(n)$ is used for $G_M(n)$.

The efficient calculation of the normalisation constant allows the evaluation of the probabilities of individual states of the network. However, for the purposes of performance analysis, one is usually *not* interested in individual state probabilities, but in mean queue lengths, mean waiting times, throughput of nodes, and utilisations. These are defined in terms of the individual state probabilities. For example, the utilisation of node i is the probability that it is not idle. This is the sum of all the state probabilities in which n_i is non-zero. Unless efficient algorithms for the evaluation of these metrics are also available, then even though the normalisation constant is known, finding performance measures will fail because of the large state space mentioned above. Fortunately, all performance measures that are of interest can be calculated efficiently as a side effect of calculating G. Suppose that we wish to evaluate the utilisation of node i; that is the probability that node i is non-empty. Denote the utilisation of node i when the network population is N by $U_i(N)$.

$$U_i(N) = \sum_{\substack{n_i > 0 \\ n_1 + \cdots + n_M = N}} \frac{1}{G} \prod_{j=1}^{M} \rho_j^{n_j}$$

Now $\gamma_i(z) - 1$ corresponds to the set of coefficients when node i is non-empty. Hence $\mathcal{U}_i(z) \triangleq \mathcal{G}_M(z)(\gamma_i(z) - 1)/\gamma_i(z)$ is the (improper) generating function of the states of the network in which node i has at least one job. When we are dealing with fixed rate servers, the ratio $(\gamma_i(z)-1)/\gamma_i(z)$ reduces to $\rho_i z$, so the coefficient of z^N, which will give the required probability is $\rho_i G_M(N-1)$ and we find that:

$$U_i(N) = \frac{\rho_i G_M(N-1)}{G_M(N)} \tag{11.4}$$

Thus in a network of fixed rate servers, the ratio of the utilisations at different servers is always constant and equal to the ratio of their respective ρ_i. The actual values of the utilisations will increase as the population of jobs increases. Some authors call the ρ values *relative utilisations*, although this is only strictly the case when all nodes are load independent.

Now the utilisation of a server can never exceed 1, so the node with the largest ρ will achieve utilisation arbitrarily close to 1 as the population increases, but the other nodes will all be limited by the ratio of their ρ_j to that with the maximum ρ. The node with the largest ρ is the bottleneck in the system and limits the system throughput.

The utilisation formula given above is just a special case of the formula for $\Pr(n_i \geq j)$, the probability that the queue length at node i is at least j. We constrain j jobs to be at node i, and allow the remaining $N - j$ to be anywhere in the network, including at node i.

$$\Pr(n_i \geq j) = \rho_i^j \frac{G_M(N-j)}{G_M(N)}$$

The marginal probability that node i has n jobs can be found in a similar way, and we find that:

$$\Pr(\text{queue } i \text{ length } m) \stackrel{\triangle}{=} p_i(m)$$

and:

$$
\begin{aligned}
p_i(m) &= \sum_{\substack{n_i = m \\ n_1 + \cdots + n_M = N}} \frac{1}{G} \prod_{j=1}^{M} \rho_j^{n_j} \\
&= \sum_{\substack{n_i \geq m \\ n_1 + \cdots + n_M = N}} \frac{1}{G} \prod_{j=1}^{M} \rho_j^{n_j} \\
&\quad - \sum_{\substack{n_i \geq m+1 \\ n_1 + \cdots + n_M = N}} \frac{1}{G} \prod_{j=1}^{M} \rho_j^{n_j} \\
&= \frac{\rho_i^m}{G(N)} \left(G(N-m) - \rho_i G(N-m-1) \right)
\end{aligned}
$$

The mean queue length at node i is found using $p_i(n)$, and the definition of the expectation. Denoting the mean number of jobs at node i when the network population is N by $N_i(N)$,

$$
\begin{aligned}
N_i(N) &= E[n_i] \\
&= \sum_{j=1}^{N} j p_i(j) \\
&= \sum_{j=1}^{N} j (\Pr(n_i \geq j) - \Pr(n_i > j)) \\
&= \sum_{j=1}^{N} j (\Pr(n_i \geq j) - \Pr(n_i \geq j+1)) \\
&= (\Pr(n_i \geq 1) - \Pr(n_i \geq 2)) \\
&\quad + 2 (\Pr(n_i \geq 2) - \Pr(n_i \geq 3)) \\
&\quad + 3 (\Pr(n_i \geq 3) - \Pr(n_i \geq 4)) + \cdots \\
&= \sum_{j=1}^{N} \Pr(n_i \geq j) - N \Pr(n_i \geq N+1)
\end{aligned}
$$

Now when the population of the network is N, $\Pr(n_i \geq N+1)$ is identically zero, so:

$$
E[n_i] = \sum_{j=1}^{N} \frac{\rho_i^j G(N-j)}{G(N)}
$$

For fixed rate servers, the mean queue length can also be evaluated by performing another convolution step for the node in question. Consider the (improper) generating functions:

$$
\begin{aligned}
\mathcal{G}^{[i]}(z) &= \mathcal{G}_M(z) \cdot \gamma_i(z) \\
\mathcal{G}^{(i)}(z) &= \mathcal{G}_M(z)/\gamma_i(z)
\end{aligned}
$$

$\mathcal{G}^{[i]}(z)$ is the generating function when node i has been included in the convolution twice; $\mathcal{G}^{(i)}(z)$ is the generating function which omits node i completely from the convolution. When node i is a fixed rate server, $\gamma_i(z) = \sum_{k=0}^{\infty} \rho_i^k z^k$, and:

$$
\begin{aligned}
\mathcal{G}^{[i]}(z) &= \mathcal{G}^{(i)}(z) \cdot \gamma_i(z) \cdot \gamma_i(z) \\
&= \mathcal{G}^{(i)}(z) \left(\sum_{k=0}^{\infty} (k+1) \rho_i^k z^k \right)
\end{aligned}
$$

Multiplying both sides by $\rho_i z$,

$$\rho_i z \mathcal{G}^{[i]}(z) = \mathcal{G}^{(i)}(z) \left(\sum_{k=1}^{\infty} k \rho_i^k z^k \right)$$

In $\mathcal{G}_M(z)$, the coefficient of z^N corresponds to the sum of all the appropriate sums of products. The coefficient of z^N in this polynomial corresponds to the sum of products, with the contribution from node i being scaled by the number of jobs there. The coefficients of $\mathcal{G}^{[i]}(z)$ can be calculated using a further round of the convolution algorithm. Define:

$$G^{[i]}(0) = 1$$
$$G^{[i]}(n) = G(n) + \rho_i G^{[i]}(n-1), \quad n > 0$$

The mean queue length is found from the z^{N-1} coefficient of $\mathcal{G}^{[i]}(z)$, multiplied by ρ_i, and normalised to give:

$$N_i(N) = \rho_i \frac{G^{[i]}(N-1)}{G(N)}$$

The throughput of node i is the rate at which jobs leave the node. Since we are only considering fixed rate servers, $X_i(N)$, the throughput of node i when the population is N, is given by:

$$\begin{aligned} X_i(N) &= \mu_i U_i(N) \\ &= \mu_i \rho_i \frac{G(N-1)}{G(N)} \\ &= v_i \frac{G(N-1)}{G(N)} \end{aligned}$$

The mean waiting time at node i, $T_i(N)$, is found using Little's theorem. The mean queue length, $N_i(N)$, is known. The mean arrival rate is equal to the mean departure rate in steady state, so it follows that:

$$T_i = N_i/X_i = \frac{\sum_{j=1}^{N} \rho^j G(N-j)}{v_i G(N-1)}$$

This algorithm, which was originally developed by Buzen, is known as the *convolution algorithm*. Its implementation as a computer program to give mean queue lengths, utilisations, etc. for a fixed rate server queueing network is a straightforward exercise. When servers whose speed depends on the queue length, or server per job nodes are included, the algorithm becomes more complex, but not significantly so.

If the speed of server i is $s_i(j)$ when there are j jobs at the server, then the term corresponding to node i in the steady state probability distribution is not $\rho_i^{n_i}$, but $\rho_i^{n_i} / \prod_{k=1}^{n_i} s_i(k)$. The improper generating functions $\gamma_i(z)$ must be altered. If node i has a load dependent server, then the calculation of $\mathcal{G}_i(z)$ from $\mathcal{G}_{i-1}(z)$ using equation (11.2), does not simplify, and must be carried out as a convolution of the two series of coefficients, as suggested by the name of the algorithm. The derivation of the utilisation of node i is identical, but the term $(\gamma_i(z) - 1)/\gamma_i(z)$ does not simplify. Writing $h_i(z)$ for $1/\gamma_i(z)$, we find that the coefficients of $h_i(z)$ can be simply calculated:

$$h_i(z) \triangleq \sum_{j=0}^{\infty} H_i(j) z^j$$

$$H_i(0) = 1$$

$$\sum_{k=0}^{j} H_i(k) f_i(j - k) = 0$$

$$H_i(j) = -\sum_{k=0}^{j-1} H_i(k) f_i(j - k)$$

The improper generating function that we are seeking is $\mathcal{U}_i(z) = \mathcal{G}_M(z) \cdot [1 - h_i(z)]$, whose jth coefficient is given by:

$$U_i(j) = G_M(j) - \sum_{k=0}^{j} G_M(k) H_i(j - k)$$

The utilisation of node i is then given by $U_i(N)/G_M(N)$. The marginal probabilities at a load dependent node can be evaluated by summing over all states that have the appropriate number of jobs at the node:

$$p_i(j) = \frac{1}{G(N)} \sum_{\substack{n_1 + \cdots + n_M = N \\ n_i = j}} \prod_{k=1}^{M} f_k(n_k)$$

Now $f_i(j)$ is a common factor, and when it is factored out, we are left with a summation which is identical to that which would arise in solving a network with no node i, the same values for v_j and μ_j at the other nodes, and with $N - j$ jobs circulating! Hence:

$$p_i(j) = f_i(j) \frac{G^{(i)}(N - j)}{G(N)} \tag{11.5}$$

where $G^{(i)}(k)$ is the normalising constant for the same network with node i 're-moved' and a population of k jobs. This normalising constant can be obtained using exactly the same algorithm. The mean queue length at a load dependent node is calculated using equation (11.5), and the definition of the mean of a distribution:

$$E[n_i] = \sum_{j=1}^{N} j p_i(j) = \sum_{j=1}^{N} j f_i(j) \frac{G^{(i)}(N-j)}{G(N)}$$

The throughput of node i, that is the number of jobs processed in unit time, can also be calculated using (11.5).

$$
\begin{aligned}
X_i(N) &= \sum_{j=1}^{N} s_i(j) \cdot \mu_i p_i(j) \\
&= \sum_{j=1}^{N} s_i(j) \cdot \mu_i f_i(j) \frac{G^{(i)}(N-j)}{G(N)}
\end{aligned}
$$

Now $s_i(j)\mu_i$ cancels with terms in the definition of $f_i(j)$, so that:

$$
\begin{aligned}
X_i(N) &= \sum_{j=1}^{N} v_i f_i(j-1) \frac{G^{(i)}(N-j)}{G(N)} \\
&= \sum_{j=0}^{N-1} v_i f_i(j) \frac{G^{(i)}(N-1-j)}{G(N)}
\end{aligned}
$$

Taking out the common factor of $v_i/G(N)$, the summation is just the discrete convolution of $f_i(j)$ and $G^{(i)}(N-1-j)$. This is exactly the last stage of the convolution algorithm for calculating $G(N-1)$, so:

$$X_i(N) = v_i \frac{G(N-1)}{G(N)} \tag{11.6}$$

So the throughput of node i depends only on the visit ratio of node i, and the ratio of the normalisation coefficients for two successive populations. It does not depend on whether the node is load-dependent or not.

11.2   Mean value analysis

Although the convolution algorithm was the first 'fast' algorithm for evaluating BCMP queueing networks, it is by no means the only one. It suffers from the disadvantage that most of the calculation effort goes into the evaluation of the normalisation constants, which have no simple interpretation in terms of the system being modelled. A number of other algorithms have been developed, each with its own characteristics and types of network for which it is best suited. *Mean value analysis* (MVA) was the first of these.

MVA depends on Little's theorem and on the so-called 'arrival theorem', which states that a job in a closed queueing network, when it enters a queue, observes the mean state of the network with itself removed. The algorithm will be described in terms of fixed rate servers. It can be modified for variable service rate nodes. Denote by $N_i(n)$, $T_i(n)$, and $X_i(n)$ the mean queue length, waiting time, and throughput of queue i when the number of jobs in the network is n. The algorithm is most easily explained if the solution to the routing equations is chosen with $v_1 = 1$. In this case, v_j can be interpreted as the mean number of visits to node j between successive visits to node 1. Now apply Little's theorem to the whole network. The boundary of the system considered includes all the nodes, and all the links between them, except that the jobs leaving node 1 leave the system and then immediately re-enter it. The mean number of jobs is fixed at n; the mean waiting time is $\sum_{j=1}^{M} v_j T_j(n)$; and the arrival rate to the system is just the rate at which jobs leave queue 1.

$$n = X_1(n) \sum_{j=1}^{M} v_j T_j(n) \tag{11.7}$$

Using the same approach that was used to develop the Pollazcek-Khintchine mean value formula in section 5.3, we consider a job arriving at node i. Applying the arrival theorem, the average queue length that it observes is the mean queue length at node i when the network population is one fewer, that is $N_i(n - 1)$. Hence its mean waiting time at node i, assuming that it is FCFS, will be:

$$T_i(n) = \frac{1}{\mu_i}(1 + N_i(n - 1)) \tag{11.8}$$

since it will wait for the jobs in front of it in the queue, and for its own service time. PS and LCFSPR nodes also satisfy the same recurrence, although a heuristic explanation is not possible. When the mean delays at all nodes have been found, the total network throughput, $X_1(n)$ can be found using (11.7). The throughput of node i is determined from $X_1(n)$, and the visit ratios:

$$X_i(n) = v_i \cdot X_1(n)$$

Finally, the mean length of the queue at node i is given by Little's theorem.

$$N_i(n) = X_i(n)T_i(n) \tag{11.9}$$

We have established a recurrence relation which expresses performance metrics for a network with n jobs in terms of the measures for the same network with $n-1$ jobs. The iteration can be started by observing that $N_i(0) \equiv 0$. Once again the complexity of the algorithm is $O(MN)$ operations. The utilisation of the server at node i is given by $X_i(n)/\mu_i$.

It is simple to include type 3, server per job nodes, in MVA. For these nodes, $T_i(n) = 1/\mu_i$, for all populations, n. For networks consisting of only load independent servers or server-per-job nodes, MVA is a simpler algorithm to implement than convolution.

If a server is load-dependent then the simple relation (11.8) does not hold. Consider a job arriving at node i and finding $k - 1$ jobs ahead of it in the queue; this occurs with probability $p_i(k - 1, n - 1)$ where $p_i(j, k)$ is the probability that node i has j jobs present when the network population is k. Before the tagged jobs leaves the node, assuming FCFS discipline, k jobs will be served at rate $\mu_i s_i(k)$, hence:

$$T_i(n) = \sum_{k=1}^{n} \frac{k}{\mu_i s_i(k)} p_i(k - 1, n - 1)$$

We need to know the marginal probabilities $p_i(k, n)$. These can also be evaluated from the marginal probabilities for the network with one fewer job. Using the formula used to calculate marginal probabilities from the convolution algorithm:

$$
\begin{aligned}
p_i(k, n) &= f_i(k) \frac{G^{(i)}(n - k)}{G(n)} \\
&= \frac{v_i}{\mu_i s_i(k)} f_i(k - 1) \frac{G^{(i)}((n - 1) - (k - 1))}{G(n)} \\
&= \frac{v_i}{\mu_i s_i(k)} \frac{G(n - 1)}{G(n)} f_i(k - 1) \frac{G^{(i)}((n - 1) - (k - 1))}{G(n - 1)} \\
&= \frac{X_i(n)}{\mu_i s_i(k)} p_i(k - 1, n - 1)
\end{aligned}
\tag{11.10}
$$

This relation, along with the obvious initial condition that $p_i(0, 0) = 1$, enables the calculation of the marginal queue length probabilities. When the marginal probabilities are known for population $n - 1$, $p_i(k, n)$ for $k = 1, 2, \ldots, n$ are found using the formula, and $p_i(0, n)$ is calculated as:

$$p_i(0, n) = 1 - \sum_{k=1}^{n} p_i(k, n) \tag{11.11}$$

It is this subtraction that makes MVA unsuitable for networks with load-dependent servers. Subtraction of nearly equal numbers is notorious in numerical analysis as a source of error. In practice, for large populations, the error becomes such that $p_i(0, n)$ can be negative!

11.3   LBANC

The algorithm known as LBANC (local balance algorithm for normalising constants) has a mixture of the properties of MVA and the convolution algorithm. The normalisation constant is calculated, as in the convolution algorithm, but the main iteration is over the network population, *not* over the nodes.

The algorithm works with *unnormalised* probabilities, which are the values of the sum of products in equation (10.7) when the normalisation constant G is 1. Unnormalised mean queue lengths $q_i(n)$ when the network population is n are defined as:

$$q_i(n) = G(n) \cdot N_i(n)$$

When node i is a load independent server, the MVA equations (11.8) and (11.9) can be combined to give a relationship between $N_i(n)$ and $N_i(n-1)$,

$$N_i(n) = U_i(n)(1 + N_i(n-1))$$

Rewriting this identity in terms of unnormalised queue lengths, and using the solution for $U_i(n)$ found in equation (11.4), the following relationship is derived,

$$\frac{q_i(n)}{G(n)} = \rho_i \frac{G(n-1)}{G(n)} \left(1 + \frac{q_i(n-1)}{G(n-1)}\right)$$
$$q_i(n) = \rho_i \left[G(n-1) + q_i(n-1)\right]$$

$G(n)$ is found from the observation that:

$$G(n) = \frac{\sum_{i=1}^{M} q_i(n)}{n}$$

since the sum of the mean queue lengths *must* equal the network population. The recurrence can be initialised by observing that $q_i(0) = 0$ and $G(0) = 1$, or equivalently $q_i(1) = \rho_i$. The complexity of LBANC when applied to the solution of networks of fixed rate nodes is $O(MN)$. Server per job nodes can be included in the LBANC algorithm, using the relationship:

$$q_i(n) = \rho_i G(n-1)$$

Once the normalisation constants have been found for each population, utilisations, and throughputs are calculated in the same way as in the convolution algorithm. Notice that the mean queue lengths are already available.

As with the other algorithms, in order to include variable rate servers in LBANC, the marginal probabilities of queue lengths must be calculated. In LBANC, they are used in unnormalised form, that is:

$$\pi_i(j, n) = \frac{p_i(j, n)}{G(n)}$$

These unnormalised probabilities satisfy a similar recurrence to that used in MVA for a load dependent server, (11.10):

$$\pi_i(j, n) = \frac{X_i(n)}{\mu_i s_i(k)} \pi_i(j-1, n-1) \quad j = 1, 2, \dots, n$$

The value of $\pi_i(0, n)$ is found in a similar manner to that used in equation (11.11):

$$\pi_i(0, n) = G(n) - \sum_{j=1}^{n} \pi_i(j, n)$$

This suffers from the same numerical problem as MVA for load-dependent servers.

11.4   Multiple class networks

All three algorithms generalise in a natural manner to multiclass networks. The network population N is replaced by a vector $\mathbf{N} = (N_1, N_2, \ldots, N_R)$ when there are R classes of job. The state of the network can be given by an $M \times R$ matrix of integers. Each row represents the jobs at the corresponding node. Columns give the distribution of jobs of that class amongst the nodes. The (i, j) element, represents the number of class j jobs at node i.

Solution of the equations for the visit ratios, (10.6), necessitates the choice of one arbitrary constant for each closed routing chain. Denote by v_{ir} the visit ratio of class r jobs to node i. For ease of explanation, networks in which jobs change class are not explicitly considered here. Allowing class change makes the terminology slightly more complex, but does not make the algorithms described below inherently more complex.

11.4.1   Convolution algorithm

When a vector-valued population $\mathbf{N}$ is used, the derivation using improper generating functions has to be expressed in terms of a vector $\mathbf{z} = (z_1, \ldots, z_R)$, with the service rate function $f_i(\mathbf{k})$ given a vector argument. The basic convolution step (11.3) is replaced by:

$$G_i(\mathbf{N}) = \sum_{\mathbf{k}=\mathbf{0}}^{\mathbf{N}} f_i(\mathbf{k}) G_{i-1}(\mathbf{N} - \mathbf{k})$$

where the sum is interpreted as:

$$\sum_{\mathbf{k}=\mathbf{0}}^{\mathbf{N}} \equiv \sum_{k_1=0}^{N_1} \sum_{k_2=0}^{N_2} \cdots \sum_{k_R=0}^{N_R}$$

As in the single class case, there are simplifications if node i is a load independent server;

$$G_i(\mathbf{n}) = G_{i-1}(\mathbf{n}) + \sum_{r=1}^{R} \rho_{ir} G_i(\mathbf{n} - \mathbf{e}_r)$$

The initial value of $G_0(\mathbf{0}) = 1$, and values of $G_i(\mathbf{n})$ for $\mathbf{n}$ with negative components are defined to be 0. The complexity of the algorithm is now $O(M \prod_{r=1}^{R}(N_r + 1))$, so the addition of extra job classes significantly increases the complexity even if the total number of jobs is unchanged.

The utilisations of the various nodes, mean queue lengths, throughputs, and marginal probabilities can all be evaluated in an analogous manner to that used in the single class case.

The throughput of class r jobs is:

$$X_{ir}(\mathbf{N}) = v_{ir} \frac{G(\mathbf{N} - \mathbf{e}_r)}{G(\mathbf{N})}$$

As with the single class case, this formula applies irrespective of the discipline at node i, and whether or not the server is load-dependent. The marginal probability that node i has n_1 jobs of class 1, n_2 of class 2, ..., n_R of class R, are calculated from the formula:

$$p_i(\mathbf{n}) = f_i(\mathbf{n}) \frac{G^{(i)}(\mathbf{N} - \mathbf{n})}{G(\mathbf{N})}$$

where $G^{(i)}(\mathbf{k})$ is the normalisation constant for the network with identical visit ratios, but omitting node i. Mean queue lengths can be derived from this formula and the definition of the mean. For load independent server nodes, the same formula applies as in the single class case, and a further convolution of the ith node will produce the mean queue length at the node.

11.4.2   MVA

Mean value analysis generalises in a straightforward manner to multiple job classes. The basic MVA relationship (11.8) is replaced by the following at node i,

$$T_{ir}(\mathbf{n}) = \frac{1}{\mu_{ir}}\left(1 + \sum_{s=1}^{R} N_{is}(\mathbf{n} - \mathbf{e}_r)\right)$$

where $N_{ir}(\mathbf{n})$ is the mean number of class r jobs present when the network population is $\mathbf{n}$. For simplicity of explanation, we assume that *every* class of job visits node 1, and that the visit ratios have been chosen so that $v_{1r} = 1$. If class r nodes do not visit node 1, then some other node will need to be used as the 'base' for the network throughput calculation for that class. This does not change the algorithm significantly. The Little's theorem derivation of the network throughput for class r is:

$$X_{1r}(\mathbf{n}) = \frac{n_r}{\sum_{j=1}^{M} v_{jr} T_{jr}(\mathbf{n})}$$

The throughput of class r jobs at node i is determined from $X_{1r}(\mathbf{n})$, and the visit ratios:

$$X_{ir}(\mathbf{n}) = v_{ir} \cdot X_{1r}(\mathbf{n})$$

Finally, the expected number of class r jobs at node i is given by:

$$N_{ir}(\mathbf{n}) = X_{ir}(\mathbf{n})T_{ir}(\mathbf{n})$$

It is possible to include load dependent nodes in a multiclass MVA algorithm using a similar relationship to that used in the single class case. The complexity of multiclass MVA is similar to that for multiclass convolution algorithm.

11.4.3 LBANC

The LBANC algorithm also extends in a natural manner to the case of multiple classes of jobs in a network. The unnormalised probabilities and unnormalised queue lengths form the basis of the algorithm. Letting the unnormalised number of class r jobs at node i be denoted by q_{ir},

$$q_{ir}(\mathbf{n}) = N_{ir}(\mathbf{n}) \cdot G(\mathbf{n})$$

The iteration for fixed rate servers can then be expressed as:

$$q_{ir}(\mathbf{n}) = \rho_{ir}\left(G(\mathbf{n} - \mathbf{e}_r) + \sum_{s=1}^{R} q_{is}(\mathbf{n} - \mathbf{e}_r)\right), \quad n_r > 0$$

Clearly, for any class which has a population of 0, the mean queue length, whether or not normalised, is also 0. At infinite server nodes, the corresponding relationship is:

$$q_{ir} = \rho_{ir} G(\mathbf{n} - \mathbf{e}_r), \quad n_r > 0$$

Extension to load dependent servers is possible too, but will not be considered here.

11.5 Dynamic scaling techniques

The convolution algorithm, whilst arguably the most effective when the network contains nodes giving load dependent service, can also suffer from numerical problems. The normalisation constant, G, is equal to the sum over all states of the unnormalised products of the states of the individual nodes. Since these unnormalised products contain arbitrary constants introduced in the course of the solution of the visit ratio equations, it is easy to see that overflow might be a problem, if some of the (non infinite server) nodes have $\rho_i > 1$. Similarly, if too many of the nodes have

$\rho_i < 1$, then there is the possibility that the value of G will be smaller than the smallest representable number on the computer, and underflow will occur. The behaviour of the normalisation constant in the convolution is related to the behaviour of the throughput as the population increases. Suppose that node i has limiting throughput $\xi_i = X_i(\infty)$. If $\rho_i < \xi_i$ then $G(n)$ expressed as a function of n initially increases and then decreases. If $\rho_i > \xi_i$ then $G(n)$ increases monotonically, and is unbounded. If $\rho_i = \xi_i$ then $G(n)$ increases monotonically, but is asymptotically bounded. This might lead one to expect that a choice of ρ_i such that it was equal to the (unknown) limiting throughput of node i, ξ_i, would avoid problems. Unfortunately, there is no guarantee that the limiting value of $G(n)$ will be within the range of representable numbers of the computer being used.

Consider first a single class network with fixed rate servers at each node, i.e., the original Buzen's algorithm network. We remarked above that the actual value of the ρ_i and the normalisation constant depend on the value arbitrarily chosen during the solution of the visit ratio equations. We make this explicit. Assume that v_k is 1, and write $\rho_i = \alpha v_i / \mu_i$, and $G_m(\alpha, n)$ for the normalisation constant when the arbitrary constant is α, the number of nodes considered is m, and the population is n. By inspection it is clear that:

$$G_m(\alpha, n) = \alpha^n G_m(1, n)$$

or, in general,

$$G_m(\beta, n) = \phi(\beta, \alpha, n) G_m(\alpha, n)$$

where:

$$\phi(\beta, \alpha, n) \stackrel{\triangle}{=} \frac{\beta^n}{\alpha^n}$$

Extending this to the multiclass network, for each routing chain, we have to choose a node whose visit ratio is defined to be 1, and introduce a scaling factor, α_k, for each routing chain. Defining $\boldsymbol{\alpha} = (\alpha_1, \ldots, \alpha_R)$, $\boldsymbol{\beta} = (\beta_1, \ldots, \beta_R)$, and including the number of nodes in the network, M, the relationship becomes:

$$G_M(\boldsymbol{\beta}, \mathbf{n}) = \Phi(\boldsymbol{\beta}, \boldsymbol{\alpha}, \mathbf{n}) G_M(\boldsymbol{\alpha}, \mathbf{n}) \tag{11.12}$$

where:

$$\Phi(\boldsymbol{\beta}, \boldsymbol{\alpha}, \mathbf{n}) \stackrel{\triangle}{=} \prod_{r=1}^{R} \left(\frac{\beta_r}{\alpha_r} \right)^{n_r}$$

The standard multiclass convolution calculation step can be rewritten to include the scaling factors explicitly. For fixed rate servers, the standard equation is (11.4.1) and when we incorporate the scaling factors explicitly it becomes:

$$G_i(\boldsymbol{\alpha}, \mathbf{n}) = G_{i-1}(\boldsymbol{\alpha}, \mathbf{n}) + \sum_{r=1}^{R} \rho_{ir} G_i(\boldsymbol{\alpha}, \mathbf{n} - \mathbf{e}_r)$$

The scaling factors must agree on each side of the equation. Equation (11.12) is used to make this so, and we have:

$$G_i(\boldsymbol{\alpha}, \mathbf{n}) = \Phi(\boldsymbol{\alpha}, \boldsymbol{\beta}, \mathbf{n}) G_{i-1}(\boldsymbol{\beta}, \mathbf{n}) + \sum_{r=1}^{R} \alpha_r \frac{v_{ir}}{\mu_{ir}} \Phi(\boldsymbol{\alpha}, \boldsymbol{\gamma}, \mathbf{n} - \mathbf{e}_r) G_i(\boldsymbol{\gamma}, \mathbf{n} - \mathbf{e}_r)$$

This equation does not make it clear that the scaling factors can in fact be different for different population vectors, even when the same node is being added to the convolution. That is the scaling factors denoted by $\boldsymbol{\gamma}$ should be written as $\boldsymbol{\gamma}(\mathbf{n} - \mathbf{e}_r)$, since the scaling factors are a function of the network population vector.

This can be used in the calculation of the normalisation constant in order to keep the value within the floating point range of the computer. Whenever the value of the normalisation constant approaches the limits of the machine's representation, so that overflow or underflow becomes likely, it is rescaled using (11.12). A small amount of storage is required in order to record the scaling factors used for different values of the population. When the performance metrics such as mean queue length, server utilisation, and node throughput are calculated, it is essential that the same scaling factor is used in all the G terms. Taking the throughput of class r jobs at node i as an example,

$$X_{ir} = v_{ir} \frac{G_M(\boldsymbol{\alpha}, \mathbf{n} - \mathbf{e}_r)}{\Phi(\boldsymbol{\alpha}, \boldsymbol{\beta}, \mathbf{n}) G_M(\boldsymbol{\beta}, \mathbf{n})}$$

11.6   Tree structured convolution

The convolution algorithm for an R class queueing network has a complexity which increases exponentially with R, both in time and space. The other algorithms suffer from similar complexity increase with respect to the number of classes. When a large number of classes are involved, the problem increases in size rapidly. In communication network modelling, it is often convenient to use a separate class of jobs for each origin destination pair. This generates a large number of classes, but they will usually have the property that jobs in any particular class only visit a small number of nodes.

The conventional convolution algorithm proceeds by successively operating on arrays corresponding to the service rates at nodes $1, 2, \ldots, M$. As node m is added, the convolution operation is performed on the array of node m service rates and the array containing the results of the convolution of nodes 1 to $m - 1$.

$$G_m(n_1, n_2, \ldots, n_R) = \sum_{k_1=0}^{n_1} \sum_{k_2=0}^{n_2} \cdots \sum_{k_R=0}^{n_R} G_{m-1}(k_1, k_2, \ldots, k_R)$$
$$f_m(n_1 - k_1, n_2 - k_2, \ldots, n_R - k_R)$$

This corresponds to the convolution of $\mathcal{G}_M(z)$ being carried out in the order:

$$\mathcal{G}_m(\mathbf{z}) = \mathcal{G}_{m-1}(\mathbf{z}) \star f_m(\mathbf{z})$$

when expressed in terms of multivariate polynomials. Convolution of polynomials is an associative and commutative operation, so the computation of:

$$\mathcal{G}_M(\mathbf{z}) = f_1(\mathbf{z}) \star f_2(\mathbf{z}) \star \cdots \star f_M(\mathbf{z})$$

can be performed in any convenient order. Having made the observation, let us for the moment consider only the 'standard' order of adding nodes to the convolution, i.e. from $i = 1$ to $i = M$.

Consider an arbitrary node, j say, which is never visited by jobs of class j. The visit ratio, v_{js}, is 0, and the value of $f_j(\mathbf{n})$ is also 0 for all vectors, $\mathbf{n}$, with $n_s > 0$. Hence, if class s jobs never visit nodes m, $m + 1$, $\ldots$, M, then after nodes 1, $\ldots$, $m - 1$ have been convolved, those elements of $G_m(\mathbf{n})$ corresponding to $n_s < N_s$ are never required since the values with which they would be convolved are all 0. In general, if a subset of nodes is *never* visited by class s jobs, then the only population vectors that need to be considered when convolving nodes within the subset are those with class s empty. Conversely, if a subset of nodes contains *all* the nodes visited by class s, the only population vectors that need to be considered for further convolution are those with $n_s = N_s$.

A class of jobs, s say, has a corresponding set of nodes, $\mathcal{S}(s)$, which are those nodes which jobs of class s visit.

$$\mathcal{S}(s) = \{i | v_{is} \neq 0\}$$

A set of nodes, $\mathcal{N}$, is said to *fully cover* a class of jobs if jobs of that class never visit nodes outside the set $\mathcal{N}$. The class is *partially covered* if the jobs visit nodes which are members of $\mathcal{N}$ and nodes which are not. The class is said to be *non-covered* if the class visits *no* nodes in $\mathcal{N}$.

$$\mathcal{S}(s) \subseteq \mathcal{N} \quad s \text{ is fully covered by } \mathcal{N}$$
$$\mathcal{S}(s) \cap \mathcal{N} = \emptyset \quad s \text{ is non-covered by } \mathcal{N}$$
$$\mathcal{S}(s) \cap \mathcal{N} \neq \emptyset \wedge \mathcal{S}(s) \nsubseteq \mathcal{N} \quad s \text{ is partially covered by } \mathcal{N}$$

With respect to this set of nodes, $\mathcal{N}$, sets of job classes can be defined by:

$$\sigma_n(\mathcal{N}) = \{r | r \text{ is non-covered by } \mathcal{N}\}$$
$$\sigma_p(\mathcal{N}) = \{r | r \text{ is partially covered by } \mathcal{N}\}$$
$$\sigma_f(\mathcal{N}) = \{r | r \text{ is fully covered by } \mathcal{N}\}$$

Assuming that all the convolutions have been done for nodes within $\mathcal{N}$, then the values which will be needed for further convolutions are those corresponding to population vectors $\mathbf{k}$:

$$\{\mathbf{k} = (k_1, k_2, \ldots, k_R) \mid k_s = 0 \text{ if } s \in \sigma_n(\mathcal{N})$$
$$k_s = N_s \text{ if } s \in \sigma_f(\mathcal{N})$$
$$k_s = 0, \ldots, N_s \text{ if } s \in \sigma_p(\mathcal{N})\}$$

Initially the tree structured algorithm starts with each node forming a set by itself. The array corresponding to the coefficients of convolution need only be calculated for the population vectors specified above. The algorithm then performs convolutions between these subsets to form a smaller number of subsets. For example, with a five node network, the subsets formed might be $\{1, 2\}$, $\{3, 5\}$, and $\{4\}$. The algorithm now has to choose which of these subsets to convolve together. It might choose to form a new subset $\{1, 2, 4\}$, and then convolve that with the $\{3, 5\}$ subset to achieve the final result. The choice of which subsets to construct and in which order has a large impact on the performance of the algorithm. Many algorithms are possible but an effective one is to build a balanced binary tree of subsets in the following order. Define the weight of a subset of nodes as:

$$w(\mathcal{N}) = \sum_{s \in \sigma_p(\mathcal{N})} |\mathcal{S}(s) \setminus \mathcal{N}|$$

i.e., the sum of the number of nodes outside $\mathcal{N}$ which are visited by jobs of classes which are partially covered by $\mathcal{N}$. The heaviest subset, $\mathcal{A}$ say, is then convolved with the subset for which the cost of convolution is least, $\mathcal{B}$ say. The cost of the convolution is measured by considering the partially covered classes of subset $\mathcal{B}$. If a class is not covered by $\mathcal{A}$, it adds 1 to the cost, if it is partially covered by $\mathcal{A}$, and fully covered by $\mathcal{A} \cup \mathcal{B}$ then -2 is added to the cost; if it is partially covered by $\mathcal{A}$, and partially covered by $\mathcal{A} \cup \mathcal{B}$, then -1 is added to the cost.

11.7 Augmented MVA

The problems of MVA when queue dependent servers are considered were hinted at in the section in which MVA was introduced. These problems can be avoided at the cost of some added computation. Instead of using equation (11.11) to calculate $p_i(0, n)$ when i is a queue dependent server, the relationship:

$$p_i(0, n) = \frac{X_j(n)}{X_j^{(i)}(n)} p_i(0, n-1)$$

is used for a single class network. In this formula j is any node except i, and $X_j^{(i)}(n)$ is the throughput of node j for the same network with node i removed. This can be demonstrated using the formula for the throughput of node j, (11.6), and the formula for the queue length distribution at node i, (11.5). Consider the ratio of $p_i(0, n)$ to $p_i(0, n - 1)$:

$$
\begin{aligned}
\frac{p_i(0, n)}{p_i(0, n - 1)} &= \frac{f_i(0)G^{(i)}(n)/G(n)}{f_i(0)G^{(i)}(n - 1)/G(n - 1)} \\
&= \frac{G^{(i)}(n)}{G^{(i)}(n - 1)} \cdot \frac{G(n - 1)}{G(n)} \cdot \frac{v_j}{v_j} \\
&= \frac{X_j(n)}{X_j^{(i)}(n)}
\end{aligned}
$$

In order to find the values of $X_j^{(i)}(n)$, the network is solved by applying the MVA algorithm, but ignoring the existence of node i. An easy way to do this is to set v_i to 0, and leave the other visit ratios unchanged. In a multiclass network, the relationship holds when the throughputs used are those of some arbitrary class k, and the population is reduced by one job from class k.

In order to solve a network with m queue dependent nodes accurately, it is necessary to apply the MVA algorithm 2^m times. First, the network is solved with all queue dependent nodes removed, and the throughputs are saved. Next, the m networks formed by considering one queue dependent node and all the load independent nodes are solved using the throughputs from stage 1. The networks consisting of two of the queue dependent nodes are solved next; there are $\binom{m}{2}$ of them. Notice that to solve the network with nodes a and b both queue dependent, we need the throughputs from the network with a removed, but including b, in order to calculate the empty probability at node a, *and* the throughputs from the network with b removed, but including a, in order to calculate the empty probability at node b. We need the throughputs of all possible combinations of j load dependent nodes, of which there are $\binom{m}{j}$, in order to calculate the throughputs in the $\binom{m}{j+1}$ different networks with $j + 1$ load dependent nodes. Thus the number of times that the MVA algorithm needs to be executed is:

$$
1 + m + \binom{m}{2} + \binom{m}{3} + \cdots + \binom{m}{m - 1} + 1 = 2^m
$$

There is also a substantial storage overhead involved in saving the values of the throughputs of the reduced networks ($2^m - 1$ of them) at all network populations of interest.

11.8 RECAL

The definition of the normalisation constant, G, as a sum of product terms, which is implied by equation (10.7), allows many interesting relationships to be drawn between the values of G corresponding to networks with different populations of jobs and service nodes. The Recursion by Chain (RECAL) algorithm is yet another algorithm which solves BCMP networks for their performance characteristics. It relates the normalisation constant of a network with r classes to the normalisation constant of one with $r - 1$ classes.

Consider a network which only has fixed rate, load independent servers, and possibly infinite server nodes. A dual network is constructed, with each load independent server node replaced by a load dependent server. At node i, the service rate in the dual network is given by $s_i(n) = n/(n + c_i)$, when n is the queue length of all classes at node i. c_i is a parameter of the dual network's node i. Denote the normalisation constant for the dual network by $G_R(\mathbf{c})$, where R is the number of different job classes, and $\mathbf{c} = (c_1, c_2, \ldots, c_M)$. Clearly, the normalisation constant of the original network is equal to $G_R(\mathbf{0})$. The dual network acts like a load independent network in which node i has an imaginary class of jobs associated with it, which only visit node i. The population of this imaginary class is given by c_i. The basic theorem of the RECAL algorithm is that $G_R(\mathbf{0})$ can be calculated using a recurrence which relates $G_r(\mathbf{c})$ and $G_{r-1}(\mathbf{c})$,

$$G_r(\mathbf{x}_r) = \sum_{\mathbf{y} \in \mathcal{Y}_r} g_r(\mathbf{x}_r, \mathbf{y}) G_{r-1}(\mathbf{x}_r + \mathbf{y}), \quad \text{for } 1 \le r \le R, \mathbf{x}_r \in \mathcal{X}_r \quad (11.13)$$

where $\mathbf{x}_r = (x_{1r}, \ldots, x_{Mr})$, $\mathbf{y} = (y_1, \ldots, y_M)$. The validity of the recurrence is the set of vectors $\mathcal{X}_r = \left\{ \mathbf{x}_r | x_{ir} \ge 0 \text{ for } 1 \le i \le M; \sum_{i=1}^{M} x_{ir} = \sum_{s=r+1}^{R} N_s \right\}$ for $1 \le r < R$, and the zero vector when $r = R$. That is, the vector $\mathbf{x}$ has non-negative components, which have their sum equal to the sum of the populations of the job classes of the original network which are excluded from this sub-network. The vector $\mathbf{y}$ ranges over the set $\mathcal{Y}_r = \left\{ \mathbf{y} | y_i \ge 0, \text{ for } 1 \le i \le M; \sum_{i=1}^{M} y_i = N_r \right\}$. In words, the values of the vector $\mathbf{y}$ are just those of the possible assignments of class r jobs to the nodes of the network. The function $g_r(\mathbf{x}_r, \mathbf{y}) = \prod_{i=1}^{M} h_{ir}(\mathbf{x}_r, \mathbf{y})$, where:

$$h_{ir}(\mathbf{x}_r, \mathbf{y}) = \begin{cases} \binom{y_i + x_{ir}}{x_{ir}} \rho_{ir}^{y_i} & \text{if node } i \text{ is FCFS, PS, or LCFSPR} \\ \frac{\rho_{ir}^{y_i}}{y_i!} & \text{if node } i \text{ is IS} \end{cases}$$

Initial conditions are given by $G_0(\mathbf{x}_0) = 1$, for all $\mathbf{x}_0 \in \mathcal{X}_0$.

The proof of this follows from explicit consideration of the normalisation constant of the dual network. Without loss of generality, nodes 1 to m are assumed to be fixed rate single server nodes in the original network, and nodes $m + 1$ to M

to be infinite server nodes. In the dual network, at nodes 1 to m, the service rate is $n/(n+c_i)$ when there are n jobs at node i. $\mathbf{n}^{(r)}$ is used to denote a state of the system, restricted to classes 1 to r. It is an $M \times r$ matrix. The ith row is the population of jobs at node i, and is denoted by $\mathbf{n}_i^{(r)}$. The total number of jobs at node i is denoted by $n_i^{(r)}$. The number of jobs of class s at node i is $n_{is}^{(r)}$. The definition of the normalisation constant, $G_r(\mathbf{c})$, is then given by:

$$G_r(\mathbf{c}) = \sum_{\mathbf{n}^{(r)} \in \mathcal{S}^{(r)}} \prod_{i=1}^{M} f_i(\mathbf{n}_i^{(r)}) \tag{11.14}$$

where $\mathcal{S}^{(r)} = \left\{ \mathbf{n}^{(r)} \mid 0 \leq n_{is}^{(r)} \leq N_s, 1 \leq s \leq r; \sum_{i=1}^{M} n_{is}^{(r)} = N_s \right\}$, and the functions $f_i(\mathbf{n}_i^{(r)})$ depend on whether or not node i is infinite server, or load independent (in the original network).

$$f_i(\mathbf{n}_i^{(r)}) = \begin{cases} \prod_{j=1}^{n_i^{(r)}} (j + c_i) \prod_{s=1}^{r} \frac{\rho_{is}^{n_{is}^{(r)}}}{n_{is}^{(r)}!} & \text{node } i \text{ is FCFS, LCFSPR, or PS} \\[2em] \prod_{s=1}^{r} \frac{\rho_{is}^{n_{is}^{(r)}}}{n_{is}^{(r)}!} & \text{node } i \text{ is IS} \end{cases}$$

Now:

$$\mathcal{S}^{(r)} = \bigcup_{\mathbf{y} \in \mathcal{Y}^{(r)}} \mathcal{S}^{(r)}(\mathbf{y})$$

where:

$$\mathcal{S}^{(r)}(\mathbf{y}) = \left\{ \mathbf{n}^{(r)} \mid \mathbf{n}^{(r)} \in \mathcal{S}^{(r)} \text{ and } n_{ir}^{(r)} = y_i \text{ for } 1 \leq i \leq M \right\}$$

That is the set $\mathcal{S}^{(r)}(\mathbf{y})$ is the set of feasible $\mathbf{n}^{(r)}$ vectors which have a total of y_i jobs at node i. Rewriting the definition of the normalisation constant, (11.14) above:

$$G_r(\mathbf{c}) = \sum_{\mathbf{y} \in \mathcal{Y}_r} \sum_{\mathbf{n}^{(r)} \in \mathcal{S}^{(r)}(\mathbf{y})} \prod_{i=1}^{M} f_i(\mathbf{n}_i^{(r)}) \tag{11.15}$$

Consider $f_i(\mathbf{n}_i^{(r)})$ for an originally load independent node, i. It contains a term of the form $\prod_{j=1}^{n_i}(j + c_i)$. Now:

$$\prod_{j=1}^{n_i^{(r)}} (j + c_i) = \prod_{j=1}^{n_i^{(r-1)} + y_i} (j + c_i)$$

$$= (1 + c_i) \cdots (y_i + c_i)(1 + y_i + c_i) \cdots (n_i^{(r-1)} + y_i + c_i)$$

$$= \frac{(y_i + c_i)!}{c_i!} \prod_{j=1}^{n_i^{(r-1)}} (j + y_i + c_i)$$

Hence, for $\mathbf{n}^{(r)} \in \mathcal{S}^{(r)}(\mathbf{y})$:

$$
\begin{aligned}
f_i(\mathbf{n}_i^{(r)}) &= \prod_{j=1}^{n_i^{(r)}} (j + c_i) \cdot \prod_{s=1}^{r} \frac{\rho_{is}^{n_{is}^{(r)}}}{n_{is}^{(r)}!} \\
&= \binom{y_i + c_i}{c_i} \prod_{j=1}^{n_i^{(r-1)}} (j + c_i) \cdot \prod_{s=1}^{r} \frac{\rho_{is}^{n_{is}^{(r-1)}}}{n_{is}^{(r-1)}!} \\
&= h_{ir}(\mathbf{c}, \mathbf{y}) \cdot \hat{f}_i(\mathbf{n}_i^{(r-1)})
\end{aligned}
$$

where $\hat{f}_i(\mathbf{n}_i^{(r-1)})$ is the service function when node i's parameter is $c_i + y_i$. Similarly, for an infinite server node:

$$
f_i(\mathbf{n}_i^{(r)}) = \frac{\rho_{ir}^{y_i}}{y_i!} \hat{f}_i(\mathbf{n}_i^{(r-1)}) = h_{ir}(\mathbf{c}, \mathbf{y}) \cdot \hat{f}_i(\mathbf{n}_i^{(r-1)})
$$

After substituting these formulas into (11.15), and taking common factors the recurrence (11.13) follows.

This general form of recurrence has not been used to calculate performance measures because there seems no simple way to relate the performance of the original network to the dual network, except by calculating all the needed normalisation constants of the original network. However, a special case of the recurrence does have convenient properties.

Consider the special case of the recurrence (11.13) when the population of each class of job is 1, that is, $N_r = 1$, for $1 \le r \le R$.

$$
G_r(\mathbf{x}_r) = \sum_{i=1}^{M} (1 + x_{ir}\theta_i)\rho_{ir} G_{r-1}(\mathbf{x}_r + \mathbf{e}_i), \quad \text{for } 1 \le r \le R, \mathbf{x}_r \in \mathcal{S}^{(r)} \quad (11.16)
$$

θ_i is an indicator function, 0 if node i is an infinite server node, 1 otherwise. Initial conditions are given by $G_0(\mathbf{x}_0) = 1$, for all $\mathbf{x} \in \mathcal{S}^{(0)}$. $\mathcal{S}^{(0)}$ is just the set of M vectors of non-negative integers whose components sum to R. Without proof, we give the probability distribution of class R jobs among the nodes of the network.

$$
\Pr(\mathbf{n}_R = \mathbf{k}) = \frac{G_{R-1}(\mathbf{k})}{G_R(\mathbf{0})} \prod_{i=1}^{M} b_i(k_i)
$$

where $\mathbf{n}_R$ is a non-negative M-vector representing the number of class R jobs at each node, $\mathbf{k}$ is a non-negative M-vector of integers, and:

$$
b_i(k_i) = \begin{cases} \rho_{iR}^{k_i} & \text{node } i \text{ is FCFS, LCFSPR, or PS} \\ \frac{\rho_{iR}^{k_i}}{k_i!} & \text{node } i \text{ is IS} \end{cases}
$$

One can also show that when $N_R = 1$, and there are *no* infinite server nodes, then:

$$G_{R-1}(\mathbf{0}) = \frac{1}{M + \sum_{r=1}^{R} N_r - 1} \sum_{i=1}^{M} G_{R-1}(\mathbf{e}_i)$$

When there are infinite server nodes, and j is an arbitrary one of them,

$$G_{R-1}(\mathbf{0}) = G_{R-1}(\mathbf{e}_j)$$

Continuing to restrict ourselves to the case of $N_R = 1$, the throughput of class R jobs at node i is given by:

$$X_{iR}(\mathbf{N}) = v_{iR} \frac{G(\mathbf{N} - \mathbf{e}_R)}{G(\mathbf{N})}$$

where the G are the normalisation constants of the original network. Since $N_r = 1$, $\mathbf{N} - \mathbf{e}_R$ has a zero component and $G(\mathbf{N} - \mathbf{e}_R) = G_{R-1}(\mathbf{0})$. Hence in terms of the dual network normalisation constants,

$$X_{iR}(\mathbf{N}) = v_{iR} \frac{G_{R-1}(\mathbf{0})}{G_R(\mathbf{0})}$$

Similarly, the mean number of class R jobs at node i, actually just the probability that the single class R job is at node i, is given by:

$$N_{iR}(\mathbf{N}) = \rho_{iR} \frac{G_{R-1}(\mathbf{e}_i)}{G_R(\mathbf{0})}$$

Utilisations and waiting times easily follow.

How can this method of finding the performance of a network, which is restricted to those networks in which the last class has only a single job in it, and the performance measures can only be calculated for this last class, be of any use for the general problem of network performance? There are two key observations to make. The first is that although usually different classes are expected to have different behaviour somewhere in the network, there is no requirement for that in the theory. Thus we can separate the jobs into as many classes as there are jobs, making each job a unique class, with its behaviour governed by its original class. Thus we can derive the performance characteristics of a *single* job of a single class, namely the one which was (arbitrarily) given the largest class index in the new network. The second key observation is that this job is statistically indistinguishable from the other jobs that had the same class in the original network. Hence, for example, the waiting times of the jobs of the original class are all identical to the waiting time of the single job.

Let us denote quantities in the new dual network with a degree symbol, so that the equivalent network has R° classes ($R^\circ = \sum_{r=1}^{R} N_r$). This is also the total population of jobs, in both networks, so $N^\circ = N$. Each job is the unique member of its class, so there is a well defined mapping from the customer class in the equivalent network to the class in the original network. The job of class s° in the equivalent network is of class $r(s^\circ)$ in the original network. Now the normalisation constant $G^\circ_{R^\circ}(\mathbf{0})$ can be found using (11.16):

$$G^\circ_s(\mathbf{x}_s) = \sum_{i=1}^{M}(1 + x_{is}\theta_i)\rho_{ir(s)}G^\circ_{s-1}(\mathbf{x}_s + \mathbf{e}_i), \quad \text{for } 1 \le s \le R^\circ$$

The vector $\mathbf{x}_s$ ranges over the set of non-negative integer M-vectors, with sum of components satisfying $\sum_{i=1}^{M} x_{is} = N^\circ - s$. Initial conditions are given by $G^\circ_0(\mathbf{x}_0) = 1$ for all $\mathbf{x}$ such that $\sum_{i=1}^{M} x_{i0} = N^\circ$. These normalisation constants are, in general, not equal to those for the original R class network, but performance measures can still be found. If the throughput of class R° jobs in the equivalent network is $X^\circ_{iR^\circ}$, then the throughput of class $r(R^\circ)$ jobs in the original network satisfies:

$$X_{ir(R^\circ)} = N_{r(R^\circ)}X^\circ_{iR^\circ}$$

since all $N_{r(R^\circ)}$ jobs in class $r(R^\circ)$ are statistically equivalent to the single job in class R°. Utilisations and mean queue lengths have the same simple relationship between the two networks, and the mean waiting times are identical. Notice that the performance measures can only be found for the class that corresponds to the last class, R°, in the equivalent network. In order to find performance measures for other classes, the jobs and classes in the equivalent network must be renumbered to make a new class correspond to R°. This involves redefining the mapping $r(s^\circ)$. By careful organisation of the algorithm, it is possible to minimise the number of recalculations of $G^\circ(\mathbf{x})$ required.

The complexity of RECAL is a little complicated to derive, but the final time complexity for a network of fixed rate servers is:

$$(4M - 1)\left(\binom{N^\circ + M - 1}{M} + \binom{R + M - 1}{M + 1} + \binom{R + M - 1}{M} - 1\right)$$
$$+ R(M + 8)$$

11.9   Mixed networks

The term *mixed network* is used to describe networks in which some customer classes are closed and others are open. The numerical solution of mixed networks can be accomplished with any of the algorithms described above, with small modifications to take into account the interaction between the open and closed classes.

Rather than consider the general mixed network, first we study a very simple case to investigate how the inclusion of open classes will affect the closed classes.

This network has M nodes, and two classes of customer. Class 1 is open and consists of jobs which arrive from outside the network at rate λ, independent of network population, are served by node M, and then leave. Thus, the visit ratios for class 1 are $v_{M,1} = 1$, $v_{i,1} = 0, i = 1, \ldots, M - 1$. Class 2 is closed; the N class 2 jobs circulate freely around the network. According top the BCMP theorem, the steady state probabilities are given by:

$$\Pr(S = (\mathbf{k}_1, \mathbf{k}_2, \ldots, \mathbf{k}_M)) = \frac{1}{G} d(S) f_1(\mathbf{k}_1) f_2(\mathbf{k}_2) \ldots f_M(\mathbf{k}_M) \qquad (11.17)$$

If there are n_1 class 1 jobs in the network when the state is S, then $d(S) = \lambda^{n_1}$. The marginal probability that there are n_1 jobs of class 1 and n_2 jobs of class 2 at node M, is given by summing the steady state probabilities of all appropriate states:

$$\Pr(n_1, n_2) = \frac{1}{G} \lambda^{n_1} f_M(n_1, n_2) \sum_{k_1 + \cdots + k_{M-1} = N - n_2} f_M(k_1) \cdots f_{M-1}(k_{M-1})$$

In order to see how the calculation of the normalisation constant is affected, the type of node M must be known. First we analyse the system as if node M is FCFS (type 1), although the same analysis applies to LCFSPR and PS nodes as well. Since node M is type 1, and $v_{M1} = 1$,

$$f_M(n_1, n_2) = \frac{(n_1 + n_2)!}{n_1! n_2!} \frac{1}{\mu_{M1}^{n_1}} \left[\frac{v_{M2}}{\mu_{M2}} \right]^{n_2}$$

The sum over all assignments of $N - n_2$ class 2 jobs to nodes from 1 to $M - 1$ is a straightforward normalisation constant calculation for the network consisting of only those nodes. Denote it by $G_{M-1}(N - n_2)$. The normalising constant of the mixed network is then given by:

$$G = \sum_{n_1 = 0}^{\infty} \sum_{n_2 = 0}^{N} \lambda^{n_1} \frac{(n_1 + n_2)!}{n_1! n_2!} \frac{1}{\mu_{M1}^{n_1}} \left[\frac{v_{M2}}{\mu_{M2}} \right]^{n_2} G_{M-1}(N - n_2)$$

Let $\rho_M = \lambda / \mu_{M1}$, and change the order of summation. Provided that $\rho_M < 1$, the sum over the class 1 population has a closed form,

$$\sum_{i=0}^{\infty} \frac{(i + j)!}{i! j!} \rho^i = \frac{1}{(1 - \rho)^{j+1}}, \quad j \geq 0, \quad \rho < 1$$

This is trivial for $j = 0$, and for larger values of j is clear by considering the $j+1$-fold product of the infinite sum $\sum_{i \geq 0} \rho^i$ with itself. Hence,

$$G = \sum_{n_2 = 0}^{N} \left[\frac{v_{M2}}{\mu_{M2}} \right]^{n_2} \frac{1}{(1 - \rho_M)^{n_2 + 1}} G_{M-1}(N - n_2)$$

A single factor of $1/(1 - \rho_M)$ may be factored out, and the summation is identical to that which results from the closed network containing only class 2 customers, except that the relative utilisation of the node, v_{M2}/μ_{M2} is reduced by the factor $1/(1 - \rho_M)$.

When node M is an infinite server node, the same formula for $\Pr(n_1, n_2)$ applies, but $f_M(n_1, n_2)$ is different:

$$\Pr(n_1, n_2) = \frac{1}{G} \lambda^{n_1} f_M(n_1, n_2) \sum_{k_1 + \cdots + k_{M-1} = N - n_2} f_1(k_1) \cdots f_{M-1}(k_{M-1})$$

$$f_M(n_1, n_2) = \frac{1}{n_1!} \left[\frac{1}{\mu_{M1}} \right]^{n_1} \frac{1}{n_2!} \left[\frac{v_{M2}}{\mu_{M2}} \right]^{n_2}$$

Summing over the feasible populations to find G:

$$G = \sum_{n_1=0}^{\infty} \sum_{n_2=0}^{N} \lambda^{n_1} \frac{1}{n_1!} \frac{1}{\mu_{M1}^{n_1}} \frac{1}{n_2!} \left[\frac{v_{M2}}{\mu_{M2}} \right]^{n_2} G_{M-1}(N - n_2)$$

$$= \sum_{n_2=0}^{N} e^{\rho_M} \frac{1}{n_2!} \left[\frac{v_{M2}}{\mu_{M2}} \right]^{n_2} G_{M-1}(N - n_2)$$

Extension of this result to the general case, where the open class visits an arbitrary set of the nodes in the network, is not difficult. At each node, the utilisation of the node by open class jobs is calculated. It is interesting that the existence of a steady state solution depends on the utilisation due to open class jobs only. If the open class utilisations are feasible, then whatever the population of the closed classes, a steady state will exist. When calculating the normalisation constant, the open class is ignored, but is used to scale various quantities appropriately. Denote the open class utilisation at node i by ρ_i. When node i is infinite server, the effect is to replace $f_i(n_O, n_C)$, where n_O is the number of open class jobs and n_C the number of closed class customers, by $e^{\rho_i} f_i^{\star}(n_C)$, where $f_i^{\star}(j)$ is the function appropriate to a similar node with *no* open classes and a total of j jobs present. If node i is a fixed rate FCFS, PS or LCFSPR server, then $f_i(n_O, n_C) = (1 - \rho_i)^{(n_C+1)} f_i^{\star}(n_C)$.

As was the case with closed networks, the performance measures such as mean queue lengths, throughputs, and delays, can all be calculated using the normalisation constants for different populations and subnetworks.

MVA and LBANC can be applied to mixed networks too. We give an MVA algorithm for mixed networks, which is of interest itself, but also has been extended to deal with priority queueing nodes. We denote by $\mathcal{O}$ the set of indices corresponding to open classes, and by $\mathcal{C}$ the indices corresponding to closed classes. The mean queue lengths are initialised to zero, for all nodes i and classes $r \in \mathcal{C}$. The network is then solved as a Jackson network, ignoring the closed classes, to find the throughputs and utilisations of open class jobs at every node:

$$X_{ir} = v_{ir}\lambda_r, \quad \rho_{ir} = \frac{X_{ir}}{\mu_{ir}}, \quad \forall r \in \mathcal{O}$$

Clearly, at a node not visited by class r, $\rho_{ir} = 0$.

Once the open chain throughputs and utilisations have been found, the MVA iterations from population $\mathbf{0}$ to $\mathbf{n}$ can start. The delay at node i to class s jobs for $s \in \mathcal{C}$ is given by:

$$T_{is} = \begin{cases} \frac{1+\sum_{t \in \mathcal{C}} N_{it}(\mathbf{n}-\mathbf{e}_s)}{\mu_{is}(1-\sum_{r \in \mathcal{O}} \rho_{ir})} & i \text{ is FCFS, PS or LCFSPR, } s \in \mathcal{C} \\ \frac{1}{\mu_{is}} & i \text{ is IS, } s \in \mathcal{C} \end{cases}$$

The throughput of class s jobs at node i is:

$$X_{is}(\mathbf{n}) = v_{is} n_s \sum_{j=1}^{M} v_{js} T_{js}(\mathbf{n}) \quad s \in \mathcal{C}$$

The mean queue length comes from Little's theorem:

$$N_{is}(\mathbf{n}) = X_{is}(\mathbf{n}) T_{is}(\mathbf{n}) \quad s \in \mathcal{C}$$

Now considering the open classes, we have, using the arrival theorem:

$$T_{ir} = \begin{cases} \frac{1+\sum_{s \in \mathcal{C}} N_{is}(\mathbf{n})}{\mu_{ir}(1-\sum_{t \in \mathcal{O}} \rho_{it})} & i \text{ is FCFS, PS or LCFSPR, } r \in \mathcal{O} \\ \frac{1}{\mu_{ir}} & i \text{ is IS, } r \in \mathcal{O} \end{cases}$$

and the mean queue length of the open class jobs at node i is found using Little's theorem:

$$N_{ir}(\mathbf{n}) = X_{ir} T_{ir}(\mathbf{n}) \quad r \in \mathcal{O}$$

This procedure is repeated for successively larger population vectors, $\mathbf{n}$, until the required network population is reached.

The iteration is repeated $\prod_{s \in \mathcal{C}}(n_s + 1)$ times. The calculation of open class performances can be moved outside the loop, since they are only dependent of the open class throughputs, calculated for the Jackson network, and on the closed class queue lengths for the population in question.

11.10   Further reading

The Gordon and Newell result[11] remained of theoretical interest because of the large state space until Buzen discovered the first fast algorithm[5]. This applied only to FCFS exponential server networks. Reiser and Kobyashi[21] developed a similar algorithm for multiple class queueing networks, incorporating most of the features of BCMP networks[2]. The generating function derivation that is given here was the work of Williams and Bhandiwad[29]. Schwetman, Balbo, and Bruell also extended Buzen's convolution algorithm to accommodate the different server disciplines and load dependent service[1]. Zahorjan[30] has also developed a modified version of convolution appropriate to multiple class networks with class switching.

Mean value analysis was developed by Reiser and Lavenberg[22]. The *arrival theorem* which is essential to the proof of MVA was independently proved by Sevcik and Mitrani[24] and Reiser and Lavenberg[18]. Reiser was responsible for augmented MVA[20]. Zahorjan and Wong[31] extended MVA to deal with mixed networks and networks with population constraints. Tucci and MacNair[27] provide a practical introduction to implementation of MVA. Bruell, Balbo and Afshari[4] consolidated all these results and algorithms, allowing mixed networks with load dependent servers.

LBANC was discovered by Chandy and Sauer[6]. They also derived a number of similar algorithms, all with slightly different complexities. Lam[15] showed how each of the convolution algorithm, MVA, and LBANC was derivable from any one of the others, confirming that they were all equally powerful.

Sauer has shown that many of the extensions mentioned in the previous chapter, such as state dependent routing and state dependency on subnetwork populations can be incorporated in the algorithms without increase in complexity[23].

Lam was also responsible for the discovery of dynamic scaling[14]. Together with Lien, he also developed the tree convolution algorithm[16]. The MVA algorithm was extended to give 'tree-MVA' by Tucci and Sauer[28], and independently by Hoyme, Bruell, Afshari, and Kain[13].

The first book dealing solely with computational algorithms for BCMP networks was by Bruell and Balbo[3]. Lazowska et al[19] gave an extended account of the use of queueing network models to analyse computer systems. Lavenberg edited a collection of papers from IBM authors on similar areas[17].

RECAL was discovered by Conway and Georganas[8]. In a recent book[9] they set all the known computational algorithms for BCMP networks in a unified framework based on decomposition of the state space. The algorithms differ by the manner in which they decompose the state space into sections that can be solved simply. A similar decomposition by chain has been applied to MVA to give an algorithm known as *Mean Value Analysis by Chain* (MVAC)[7]. Yet another decomposition approach gives the *Distribution Analysis by Chain* algorithm (DAC), developed by de Souza e Silva and Lavenberg[10]. Greenberg and McKenna developed a tree version of RECAL, and implemented their algorithm on a parallel processor[12].

The accuracy of product form algorithms, even when the assumptions underlying the analysis are not satisfied, is analysed by Suri[25]. The effect of parameter estimation errors is dealt with in Suri and Tay[26].

11.11 Bibliography

[1]   G. Balbo, S.C. Bruell, and H.D. Schwetman. Customer classes and closed network models — a solution technique. In B. Gilchrist, editor, *Information Processing 77*, pages 559–564. North Holland, Amsterdam, 1977.

[2]   F. Baskett, K.M. Chandy, R.R. Muntz, and F. Palacios-Gomez. Open, closed and mixed networks of queues with different classes of customers. *Journal of the ACM*, 22(2):248–260, April 1975.

[3]   S.C. Bruell and G. Balbo. *Computational Algorithms for Closed Queueing Networks*. North Holland, Amsterdam, 1980.

[4]   S.C. Bruell, G. Balbo, and P.V. Afshari. Mean value analysis of mixed multiple class BCMP networks with load-dependent service stations. *Performance Evaluation*, 4(4):241–260, 1984.

[5]   J.P. Buzen. Computational algorithms for closed queueing networks with exponential servers. *Communications of the ACM*, 16(9):527–531, September 1973.

[6]   K.M. Chandy and C.H. Sauer. Computational algorithms for product form queueing networks. *Communications of the ACM*, 23(10):573–583, October 1980.

[7]   A.E. Conway, E. de Souza e Silva, and S.S. Lavenberg. Mean value analysis by chain of product form queueing networks. *IEEE Transactions on Computers*, C-38(3):432–442, March 1989.

[8]   A.E. Conway and N.D. Georganas. A new method for computing the normalization constant of multiple-chain queueing networks. *INFOR*, 24(3):184–198, 1986.

[9]   A.E. Conway and N.D. Georganas. *Queueing Networks — Exact Computational Algorithms*. MIT Press, Cambridge, Massachusetts, 1989.

[10] E. de Souza e Silva and S.S. Lavenberg. Calculating joint queue length distributions in product form queueing networks. *Journal of the ACM*, 36(1):194–207, January 1989.

[11] W.J. Gordon and G.F. Newell. Closed queueing networks with exponential servers. *Operations Research*, 15(2):254–265, March/April 1967.

[12] A.G. Greenberg and J. McKenna. Solution of closed, product form, queueing networks via the RECAL and tree-RECAL methods on a shared memory multiprocessor. *Performance Evaluation Review*, 17(1):127–135, 1989.

[13] K.P. Hoyme, S.C. Bruell, P.V. Afshari, and R.Y. Kain. A tree-structured mean value analysis algorithm. *ACM Transactions on Computer Systems*, 4(2):178–185, May 1986.

[14] S.S. Lam. Dynamic scaling and growth behavior of queueing network normalization constants. *Journal of the ACM*, 29(2):492–513, April 1982.

[15] S.S. Lam. A simple derivation of the MVA and LBANC algorithms from the convolution algorithm. *IEEE Transactions on Computers*, C-32(11):1062–1064, November 1983.

[16] S.S. Lam and Y.L. Lien. A tree convolution algorithm for the solution of queueing networks. *Communications of the ACM*, 26(3):203–215, March 1983.

[17] S.S. Lavenberg. *Computer Performance Modeling Handbook*. Academic Press, London, 1983.

[18] S.S. Lavenberg and M. Reiser. Stationary state probabilities at arrival instants for closed queueing networks with multiple types of customers. *Journal of Applied Probability*, 17(4):1048–1061, 1980.

[19] E.D. Lazowska, J.L. Zahorjan, G.S. Graham, and K.C. Sevcik. *Quantitative System Performance: Computer System Analysis Using Queueing Network Models*. Prentice-Hall, Englewood Cliffs, NJ, 1984.

[20] M. Reiser. Mean value analysis and convolution method for queue dependent servers in closed queueing networks. *Performance Evaluation*, 1(1):7–18, 1981.

[21] M. Reiser and H. Kobayashi. Queueing networks with multiple closed chains: Theory and computational algorithms. *IBM Journal of Research and Development*, 19(3):282–294, May 1975.

[22] M. Reiser and S.S. Lavenberg. Mean value analysis of closed multichain queueing networks. *Journal of the ACM*, 27(2):313–322, April 1980.

[23] C.H. Sauer. Computational algorithms for state-dependent queueing networks. *ACM Transactions on Computer Systems*, 1(1):67–92, February 1983.

[24] K.C. Sevcik and I. Mitrani. The distribution of queueing network states at input and output instants. *Journal of the ACM*, 28(2):358–371, April 1981.

[25] R. Suri. Robustness of queueing network formulas. *Journal of the ACM*, 30(3):564–594, July 1983.

[26] Y.C. Tay and R. Suri. Error bounds for performance prediction in queueing networks. *ACM Transactions on Computer Systems*, 3(3):227–254, August 1985.

[27] S. Tucci and E.A. MacNair. Implementation of mean value analysis for open, closed and mixed queueing networks. *Computer Performance*, 3(4):233–239, 1982.

[28] S. Tucci and C.H. Sauer. The tree MVA algorithm. *Performance Evaluation*, 5(3):187–196, 1985.

[29] A. Williams and R.A. Bhandiwad. A generating function approach to queueing network analysis of multiprogrammed computers. *Networks*, 6(1):1–22, 1976.

[30] J.L. Zahorjan. An exact solution method for the general class of closed separable queueing networks. *Performance Evaluation Review*, 8(3):107–112, 1979.

[31] J.L. Zahorjan and E. Wong. The solution of separable queueing network models using mean value analysis. *Performance Evaluation Review*, 10(3):80–85, 1981.

Chapter 12

Approximations and bounds

This chapter collects together a number of approximation techniques to solve networks which are not strictly product form using product form algorithms. They are also of use to find approximate solutions to product form networks more quickly than the exact algorithms described in the previous chapter. Closely related to approximation methods is the idea of calculating bounds on performance. Although the algorithms for exact solution of product form networks are fast and accurate, for the purposes of examining alternative designs during early phases of a system design, or for confirming or denying some hypothesis about system performance, bounds on system performance may provide enough information. Notice that the only justification for using bounds rather than exact results is that they are cheaper to calculate.

12.1 Approximate MVA and Linearizer

The MVA algorithm for multiple class networks is based on an iterative step that takes statistics for the network with one less customer in each class, and calculates those statistics for the population of interest. The arrival theorem gives:

$$T_{ir}(\mathbf{n}) = \frac{1}{\mu_{ir}}(1 + \sum_{s=1}^{R} N_{is}(\mathbf{n} - \mathbf{e}_r)) \tag{12.1}$$

where $N_{ir}(\mathbf{n})$ is the mean number of class r jobs present when the network population is $\mathbf{n}$. Little's theorem, and throughput considerations give:

$$N_{ir}(\mathbf{n}) = \frac{v_{ir}T_{ir}(\mathbf{n})}{\sum_{j=1}^{M} v_{jr}T_{jr}(\mathbf{n})} \cdot n_r \tag{12.2}$$

Combining equations (12.1) and (12.2) allows the mean queue lengths when the population is $\mathbf{n} - \mathbf{e}_s$ to be used to calculate the mean queue lengths when the population is $\mathbf{n}$. The classical MVA algorithm uses the population $\mathbf{0}$ to initialise the iterations. Approximate MVA algorithms use the mean queue lengths at population $\mathbf{n}$ to estimate the mean queue lengths at population $\mathbf{n} - \mathbf{e}_s$. A single iteration of MVA is then performed to give new estimates of the queue lengths at population $\mathbf{n}$. These steps can be carries out repeatedly until the mean queue lengths converge.

The first approximate MVA algorithm assumed that the removal of a single job from a network with a large population would have a rather small effect, so the estimate for $N_{ir}(\mathbf{n} - \mathbf{e}_s)$ was proportional to $N_{ir}(\mathbf{n})$:

$$N_{ir}(\mathbf{n} - \mathbf{e}_s) = \begin{cases} N_{ir}(\mathbf{n}) & \text{for } r \neq s \\ \frac{n_j - 1}{n_j} N_{ij} & \text{for } r = s \end{cases}$$

After initial estimates have been given for the mean queue lengths, this formula is used alternately with a single iteration of MVA, iterating until the successive estimates of mean queue length at each node, for each class have converged. This estimate is justified when all classes have large populations. If one or more classes of jobs have small populations, then the assumption that the removal of a job from that class will not impact the network is false. In the extreme, a class consisting of a single job that only visits a single node in the network, and that has a high service requirement will dramatically change network performance when it is removed.

An alternative estimation of the mean queue lengths at the lower population can be derived as follows. Define $\phi_{ir}(\mathbf{n})$ as the fraction of class r jobs that are at node i, on average, when the population is $\mathbf{n}$.

$$\phi_{ir}(\mathbf{n}) = \frac{N_{ir}(\mathbf{n})}{n_r}$$

$D_{irs}(\mathbf{n})$ is defined as the change in $\phi_{ir}(\mathbf{n})$ when a single class s job is removed:

$$D_{irs}(\mathbf{n}) = \phi_{ir}(\mathbf{n} - \mathbf{e}_s) - \phi_{ir}(\mathbf{n})$$

It follows that:

$$N_{ir}(\mathbf{n} - \mathbf{e}_s) = \begin{cases} n_r(\phi_{ir}(\mathbf{n}) + D_{irs}(\mathbf{n})) & \text{for } r \neq s \\ (n_r - 1)(\phi_{ir}(\mathbf{n}) + D_{irs}(\mathbf{n})) & \text{for } r = s \end{cases} \tag{12.3}$$

Notice that $\phi_{ir}(\mathbf{n})$ can be calculated, but that $D_{irs}(\mathbf{n})$ cannot, because it depends on the very parameter that we are trying to estimate, $N_{ir}(\mathbf{n} - \mathbf{e}_s)$. The basic approximation consists of starting with initial estimates of $N_{ir}(\mathbf{n})$ and $D_{irs}(\mathbf{n})$, and iterating using equation (12.3) and a single iteration of MVA until the estimates of queue lengths converge. The simple approximation provided above corresponds to $D_{irs}(\mathbf{n}) = 0$.

The Linearizer algorithm uses the above procedure as part of its operation. It assumes that $D_{irs}(\mathbf{n} - \mathbf{e}_t) = D_{irs}(\mathbf{n})$, that is that the fraction of jobs at particular nodes is a *linear* function of population. The algorithm is initialised by assuming that the jobs are uniformly distributed around the network. That is:

$$N_{ir}(\mathbf{n}) = \frac{n_r}{M} \tag{12.4}$$

$$N_{ir}(\mathbf{n} - \mathbf{e}_s) = \begin{cases} \frac{n_r}{M}, & r \neq s \\ \frac{n_r - 1}{M} & r = s \end{cases} \tag{12.5}$$

Initialise the values of the $D_{irs}(\mathbf{n})$ to 0.

The basic approximation algorithm is then applied, iterating until it converges. Any internal values generated by this algorithm must be hidden from Linearizer. For example, the basic approximation estimates $N_{ir}(\mathbf{n} - \mathbf{e}_s)$; these estimates must not be used by Linearizer. The only output from this application of the approximation is $N_{ir}(\mathbf{n})$. These estimates are compared with those available at the previous iteration, and if convergence has been reached, the algorithm terminates. If no convergence has been achieved, then the basic approximation is used to evaluate $N_{ir}(\mathbf{n} - \mathbf{e}_s)$ for all classes r and s. The estimates used for input to the basic approximation are the private variables of Linearizer, *not* any intermediate values generated by previous uses of the basic approximation. The value of $D_{irs}(\mathbf{n} - \mathbf{e}_t)$ is identical to $D_{irs}(\mathbf{n})$, this being Linearizer's approximation. The new estimates for $N_{ir}(\mathbf{n} - \mathbf{e}_s)$ are then used to calculate new estimates of $D_{irs}(\mathbf{n})$. The algorithm is then repeated.

Although one can use a convergence test to terminate the Linearizer iterations, all the practical experience that has been accumulated with use of the algorithm suggests that three iterations only are sufficient. Further iterations add very little accuracy to the estimates.

12.2 Proportional approximation method

Although only three iterations of the Linearizer algorithm are required, each of these iterations consists of many iterations of the basic MVA approximation with each of these basic approximation steps taking $O(MR^2)$ operations. The *proportional approximation method* (PAM) is a non-iterative approximation method, startling in its simplicity and remarkably accurate. Essentially, PAM replaces the uniform initialisation of queue lengths used by Linearizer with initial conditions which reflect the relative speeds and loads of the nodes.

The basic PAM algorithm initialises the mean queue lengths at each node by first estimating the proportion of jobs of class r at node i as:

$$p_{ir} = \frac{v_{ir}/\mu_{ir}}{\sum_{k=1}^{M} v_{kr}/\mu_{kr}}$$

The mean queue lengths are then estimated by:

$$N_{ir}(\mathbf{n}) = p_{ir} n_r$$

$$N_{ir}(\mathbf{n} - \mathbf{e}_s) = \begin{cases} p_{ir} n_r, & r \neq s \\ p_{ir}(n_r - 1), & r = s \end{cases}$$

This initial estimation can be compared with Linearizer's initialisations, (12.4) and (12.5). Using this initialisation, the mean delay encountered by a class r job at node i, assuming node i is a fixed rate server, is:

$$T_{ir}(\mathbf{n}) = 1/\mu_{ir}(1 + \sum_{s=1}^{R} N_{is}(\mathbf{n} - \mathbf{e}_r))$$

Of course, for an infinite server node, $T_{ir}(\mathbf{n}) = 1/\mu_{ir}$. The total network throughput of class r jobs is then given by:

$$X_{\cdot r}(\mathbf{n}) = \frac{n_r}{\sum_{i=1}^{M} v_{ir} T_{ir}(\mathbf{n})}$$

The throughput of individual nodes, either in total, or by class, can easily be found, along with new estimates of the mean queue lengths, node utilisations, etc.

This basic PAM algorithm works well, so long as none of the servers is a bottleneck. A simple improvement overcomes this drawback. The server utilisations are calculated:

$$U_i(\mathbf{n}) = \sum_{r=1}^{R} \frac{v_{ir}}{\mu_{ir}} X_{\cdot r}(\mathbf{n})$$

If all nodes i which are fixed rate servers have utilisations less than one, then no improvements can be made. However, it is possible that some servers have utilisations greater than one, since the basic PAM is an approximation. For each class r, find the server with maximum utilisation, and set:

$$Y_r = \max_i U_i(\mathbf{n})$$

where the maximum is taken over the set of nodes visited by class r. For all those classes r with $Y_r > 1$, scale the throughput of the class by:

$$X_{\cdot r}(\mathbf{n}) = \frac{X_{\cdot r}}{Y_r}$$

The node throughputs, utilisations, etc., should now be recalculated using the revised class throughputs.

These PAM algorithms are good approximations when the job classes visit relatively few nodes each, and/or the service times of jobs at each node are of the same order of magnitude at all nodes. If the service times are highly variable, or if each class of jobs visits a large number of the nodes of the network, then errors in the estimates of class throughputs can be as high as 40% if the maximum server utilisation is close to one.

12.3   Priority approximations

Queueing networks in which one class of customers has priority over another are very common in practice. For example, it is usual for scheduling algorithms to attempt to discriminate in favour of jobs that are I/O bound, in order to keep both CPU and I/O systems fully utilised. In packet switching communication networks, priority may be given to interactive transactions at the expense of file transfer and other long duration activities. Separable networks do not allow classes of customers with different priorities.

12.3.1   Reduced availability approximations

The simplest approximation for priority servers in networks is the *reduced work-rate* approximation. It applies to preemptive resume scheduling. The processing power available to class k customers is reduced by the proportion of time that the server is processing jobs of higher priority classes. If the service rate of class k customers at node j is μ_{jk}, and the utilisation of node j by class r jobs is ρ_{jr}, then the effective service rate at node j for class k jobs is:

$$\hat{\mu}_{jk} = \mu_{jk}(1 - \sum_{r=1}^{k-1} \rho_{jr})$$

The network can be solved exactly for the highest priority class of jobs. The performance of successively lower classes of jobs can then be calculated. Notice that this assumption limits reduced work-rate approximations to circumstances in which the priority structure is the same at each node in the network. At infinite server nodes the priority structure is irrelevant.

In order to allow different classes of job to have different priority at different nodes, a more complex algorithm is needed. The *shadow approximation* extends the reduced work-rate approximation to more general networks. In the shadow approximation, a node giving priority service is decomposed into P shadow nodes, where P is the number of priority classes at the original node. Each shadow node serves one priority class only, and the shadow node corresponding to node i in the original network, serving class r jobs, has service rate:

$$\hat{\mu}_{ir} = \mu_{ir}(1 - \sum_{s=1}^{r-1} \rho_{is})$$

where ρ_{is} is the utilisation of the shadow node of node i serving class s jobs. The summation is over all the classes which have priority over class r at node i. The utilisations ρ_{is} are unknown, so the approximation is initialised with estimates, calculates the service rates at the shadow nodes, and then solves the BCMP network containing the shadow nodes in order to find new values for the utilisations. This procedure is iterated until the utilisations converge.

There are a number of problem areas in the reduced availability approximation. *Delay error* is that component of the approximation error due to the fact that low priority jobs start service as soon as they arrive at the (shadow) node. In reality, low priority jobs are delayed until there are no higher priority jobs present. *Synchronisation error* is the result of the interdependency of the nodes. When a low priority job arrives at a node, it is known that there are no higher priority jobs at the node which has been left by the low priority job. The accuracy of the reduced availability approximation is also affected by interarrival time variability. When the high priority jobs have a greater mean service time than low priority jobs, the arrival theorem overestimates the queue length observed by arriving high priority jobs. A number of modifications can be made to account for these errors. For instance, the synchronisation error can be (partly) accounted for by replacing the reduced availability definition is replaced by:

$$\hat{\mu}_{ir} = \mu_{ir}(1 - p(n_1 > 0, n_2 > 0, \ldots, n_{r-1} > 0 | n_r > 0))$$

where $p(\cdot)$ is the probability associated with the number of jobs at node i. Further modifications are possible, based on an MVA algorithm, which allow service time variability to be accounted for and delay error to be reduced.

12.3.2 Delay modification approximations

Reduced availability approximations are only effective for preemptive resume priority servers. High priority jobs are completely unaffected by jobs of lower priorities.

The only approximation which can treat both preemptive and non-preemptive priorities alter the fundamental calculation of delay at the heart of the MVA iteration. From the analyses presented in chapter 7, the mean delays in $M/M/1$ priority queues are known as functions of the arrival rates, utilisations and service rates of the different priority classes. In particular, for preemptive resume priority, consider a class r job. Its mean delay is given by the sum of its own service time, the service times of all jobs of its own or higher priority classes that are present when it arrived, and the service times of those jobs of higher priority classes which arrive during its residence at the node. Hence:

$$W_r = \frac{1}{\mu_r} + \sum_{s=1}^{r} N_s \cdot \frac{1}{\mu_s} + \sum_{s=1}^{r-1} W_r \frac{\lambda_s}{\mu_s}$$

Notice that since services are assumed to be exponentially distributed, the expected residual service time is identical to the expected service time. Solving for W_r:

$$W_r = \frac{1/\mu_r + \sum_{s=1}^{r} N_s/\mu_s}{1 - \sum_{s=1}^{r-1} \rho_s} \tag{12.6}$$

where ρ_s is the utilisation by class s jobs, and N_s is the mean number of class s jobs in the queue, including the one in service.

In non-preemptive $M/M/1$ systems, as soon as the job has started service it will not be interrupted. The arrivals after the tagged job which will cause extra delay to the tagged job are those of higher priority classes that arrive in the queueing time of the tagged job. The delay due to jobs of the same or higher priority class that are present when the tagged job arrives is the same as in the preemptive case; however, the tagged job must also wait for the completion of service of any lower priority job which might be in service when the tagged job arrives. Combining these terms:

$$W_r = \frac{1}{\mu_r} + \sum_{s=1}^{r} N_s \cdot \frac{1}{\mu_s} + \sum_{s=r+1}^{R} \frac{\rho_s}{\mu_s} + \sum_{s=1}^{r-1} (W_r - \frac{1}{\mu_r}) \frac{\lambda_s}{\mu_s}$$

Rearrangement in terms of W_r gives:

$$W_r = \frac{1}{\mu_r} + \frac{\sum_{s=1}^{r} N_s/\mu_s + \sum_{s=r+1}^{R} \rho_s/\mu_s}{1 - \sum_{s=1}^{r-1} \rho_s} \tag{12.7}$$

These formulas are exact for an $M/M/1$ system. The modified delay approximation to MVA consists of using these expressions in place of the arrival theorem expression for the waiting time at population n in terms of queue lengths at population $n-1$. We shall use the mixed network formulation of MVA, given in section 11.9. The value of N_s to be substituted in (12.6) or (12.7) is the mean number of class s jobs observed by an arrival of class r. The arrival theorem, which *does not hold* in this case, states that arrivals in open classes see the steady state probabilities, and in closed classes see the steady state probabilities of the network with themselves removed. The arrival theorem does not hold for priority queueing networks, because the class of the job arriving may constrain the probabilities that If class r is open ($r \in \mathcal{O}$), then the steady state mean queue length of class s jobs is used. In this case, the class r mean queue length can be eliminated using Little's theorem. If class r is closed ($r \in \mathcal{C}$), then the steady state mean queue length when the network population has one fewer class r job is used.

$$T_{ir}(\mathbf{n}) = \begin{cases} \frac{1/\mu_{ir} + \sum_{s=1}^{r} N_{is}(\mathbf{n}-\mathbf{e}_r)/\mu_{is}}{1 - \sum_{s=1}^{r-1} \rho_{is}}, & r \in \mathcal{C} \\ \frac{1/\mu_{ir} + \sum_{s=1}^{r-1} N_{is}(\mathbf{n})/\mu_{is}}{1 - \sum_{s=1}^{r-1} \rho_{is}}, & r \in \mathcal{O} \end{cases}$$

The only unknowns remaining in this equation are the values of ρ_{is}. For open classes, ρ_{is} is given by X_{is}/μ_{is} and does not vary with the closed network population. For classes that are closed, the node utilisations by class depend on the network population. A number of network populations been tried, but the best results are achieved by using $\rho_{is}(\mathbf{n} - N_{is}(\mathbf{n}))$, that is the node utilisation by class s jobs when the population of class s jobs is reduced by the mean number of class s jobs at node i. The heuristic justification for this approximation is that if there are $N_{is}(\mathbf{n})$ jobs at node i, then the arrival rate of class s jobs at node i will be a function of the remaining jobs in the network.

12.4   Flow equivalent aggregation

In many cases, the analysis of queueing networks is used to explore the characteristics of a single node or possibly a set of nodes in a network. For example, the
response time of a transaction processing system may be investigated to discover
the best combination of speed and number of disks to employ. As far as the rest
of the system is concerned, the disk system is a 'black box' which has a (probably load-dependent) service rate. The internal organisation of the disk system only
affects the performance of the whole system through these service rates. Similarly,
although there are a number of other components in the model, e.g. CPU, terminals,
front-end processors etc., as far as the disk system is concerned, they merely provide
input requests and consume the output from the disk system.

Since a reduction in the number of nodes will reduce the time taken to solve
a queueing network model, can sets of nodes be grouped together, and their 'black
box' parameters discovered to enable the use *flow equivalent servers*? It is possible to
do this, and find the parameters of the flow equivalent server using *Norton's theorem*
for queueing networks. The idea is based on Norton's theorem for electrical circuit
analysis, which replaces a complex circuit by a black box with the same behaviour.

The construction of a flow equivalent server involves the separation of the
network nodes into two subsets, the set of nodes of interest, and the rest. Instead of
investigating the whole parameter space for the set of nodes of interest by solving the
whole network again and again, a flow equivalent server is constructed to represent
the effect on the nodes in the subset of interest by the other nodes. The model we
use as an example is similar to that described above. A number of terminals submit
requests to a transaction processing system. the requests must be processed by a front
end processor(FEP), before being passed to the main system. This consists of a CPU,
and M disks. We assume that the FEP and the disks give FCFS service, the CPU is
a processor sharing node, and the terminals can be represented as an infinite server
node. The terminals and the requests which they submit are statistically identical, so
a single class model will be used. Figure 12.1 shows the topology of the system, and
with appropriate exponential service assumptions the network is separable. Assuming load independent service, M disks, and N terminals, the normalisation constant
can be found in $O((M + 3)N)$ operations. The management desires to evaluate a
number of different FEPs. A flow equivalent server can be constructed, so that the
FEP can be examined in isolation. The modified network is shown in Figure 12.2,
where the flow equivalent server behaves in the same way as the CPU, disks and
terminals combined, when viewed from the FEP. The flow equivalent server will be
load dependent. Its parameters are calculated by finding the throughput of the FEP
in the original network when the FEP is 'short circuited', or equivalently, replaced
by a server with mean service time 0. This network is shown in Figure 12.3. The

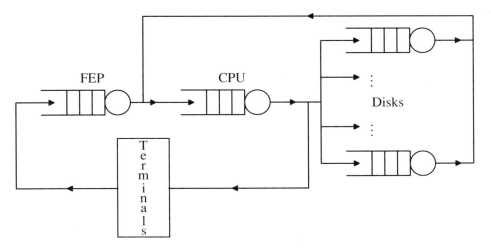

Figure 12.1 Transaction processing system

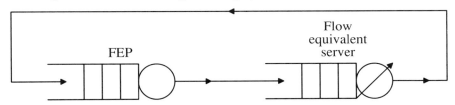

Figure 12.2 Flow equivalent server

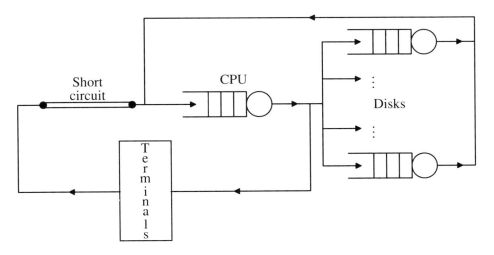

Figure 12.3 Short circuited network

throughput of this short circuit at network populations $1, 2, 3, \ldots, N$, is the service rate of the flow equivalent server when there are $1, 2, \ldots, N$ jobs at that node.

12.5   Asymptotic bound analysis

The cheapest set of bounds to calculate are the so called *asymptotic bounds*. They can be applied to networks in which all nodes give service in a load independent manner. The load at each node in the network is defined to be the the average number of times a job visits that node times the service time required per visit. In the notation of chapter 11, $L_k = v_k/\mu_k$. Assume that node b has the maximum loading, i.e., $L_b = \max_k L_k$. The node with maximum loading is often described as the *bottleneck device*. When the population is n jobs, denote the network throughput by $X(n)$, and the utilisation of node i by $U_i(n)$. The utilisation of a node is the product of the load on that node and the system throughput.

$$U_k(n) = X(n)L_k$$

By definition, The maximum utilisation that *any* node can have is 1, hence:

$$U_k(n) \leq 1$$

Combining these observations,

$$X(n) \leq \frac{1}{L_b}$$

As the system population grows, throughput can grow at best linearly. This would only be possible with perfect scheduling. Hence:

$$X(n) \leq nX(1)$$

Now the response time when the population is 1 is $R(1) = \sum_k L_k$. Hence $X(1) = 1/R(1)$. Combining these two bounds, the asymptotic bound on system throughput is given by:

$$X(n) \leq \min\left(\frac{1}{L_b}, \frac{n}{R(1)}\right)$$

The same observations can be used to express bounds on system response time, rather than on throughput. Little's theorem gives $R(n) = n/X(n)$. Hence:

$$R(n) \geq \max\left(nL_b, R(1)\right)$$

Asymptotic bounds are the extremes of system behaviour; either no queueing is taking place, or all jobs are queueing at the bottleneck device. In throughput terms, either the throughput grows linearly with network population, or the bottleneck device saturates with utilisation 1, and the throughput is limited by that device's throughput. They are extremely simple to calculate.

12.6   Balanced job bounds

Asymptotic bounds only provide very crude lower bounds on response time and upper bounds on system throughput. Although adequate in some situations, both upper and lower bounds on response time are more useful. Balanced job bounds provide a simple to calculate set of bounds. The idea behind them is to assume that all nodes are identical with load L, and calculate the mean response times, throughputs, etc. Appropriate choice of L can then provide optimistic and pessimistic bounds on system performance. Balanced job bounds apply to separable queueing networks only.

Consider a network in which all loadings are equal to L, and use MVA to evaluate its performance. The application of Little's theorem, as in equation (11.7), gives:

$$X(n) = \frac{n}{\sum_k R_k(n)} \tag{12.8}$$

where $R_k(n)$ is the total time spent at node k in all visits, both queueing and being served. The arrival theorem gives:

$$R_k(n) = L_k(1 + N_k(n-1))$$

where $N_k(n-1)$ is the mean queue length at node k when the network population is $n-1$. Since all loadings L_k are identical, and equal to L, we have:

$$\begin{aligned} \sum_k R_k(n) &= \sum_k L(1 + N_k(n-1)) \\ &= L(M+n-1) \end{aligned}$$

where M is the number of nodes in the network. Hence:

$$X(n) = \frac{n}{(M+n-1)L}$$

In order to apply the balanced job bounds to realistic networks, the node with minimum load L_m and the bottleneck node L_b must be identified. The throughput for an arbitrary separable queueing network is certainly greater than the throughput for the same network with all node loadings increased to L_b. Similarly, the throughput cannot exceed the throughput of a balanced network in which each node has load L_m. Hence:

$$\frac{n}{(M+n-1)L_b} \leq X(n) \leq \frac{n}{(M+n-1)L_m}$$

Using Little's theorem, the response time bounds are given by:

$$(M+n-1)L_m \leq R(n) \leq (M+n-1)L_b$$

These bounds can actually be improved somewhat. Consider the denominator in equation (12.8):

$$\sum_k R_k(n) = \sum_k L_k(1 + N_k(n-1)) = R(1) + \sum_k L_k N_k(n-1)$$

since $R(1)$ is simply the sum of the loadings. Since $L_b \geq L_k$,

$$\sum_k R_k(n) \leq R(1) + (n-1)L_b$$

and hence

$$X(n) \geq \frac{n}{R(1) + (n-1)L_b}$$

An upper bound can be found by considering the mean queue length at each node. From the convolution algorithm:

$$N_k(n) = \sum_{j=1}^{n} \alpha_j L_k^j$$

where the values α_j are positive coefficients which are independent of the node concerned, k. It follows that

$$L_i \geq L_j \Rightarrow N_i(n) \geq n_j(n)$$

Hence,

$$\sum_k L_k N_k(n-1) \geq L_a \sum_k N_k(n-1) = L_a(n-1)$$

where L_a is the average load on a node. If $a_1 \geq a_2 \geq \cdots a_n \geq 0$ and similarly for the terms b_i, then $\sum a_i b_i \geq \bar{a} \sum b_i$, where $\bar{a}$ is the average of the a_i. This can be proved by considering the sum in two parts, split at the kth term, with $a_k \geq \bar{a} \geq a_{k+1}$.

$$\sum_{i=1}^{n} b_i(a_i - \bar{a}) = \sum_{i=1}^{k}(b_i - b_k)(a_i - \bar{a}) + b_k \sum_{i=1}^{k}(a_i - \bar{a})$$

$$+ \sum_{i=k+1}^{n}(b_k - b_i)(\bar{a} - a_i) - b_k \sum_{i=k+1}^{n}(\bar{a} - a_i)$$

$$= \sum_{i=1}^{k}(b_i - b_k)(a_i - \bar{a}) + \sum_{i=k+1}^{n}(b_k - b_i)(\bar{a} - a_i)$$

By construction, both factors of all terms of the summations are nonnegative, so the relationship $\sum a_i b_i \geq \bar{a} \sum b_i$ follows. Replacing the sum in the denominator of (12.8), it follows that:

$$X(n) \leq \frac{n}{R(1) + (n-1)L_a}$$

By use of Little's result, the equivalent bounds on response time are:

$$R(1) + (n-1)L_a \leq R(n) \leq R(1) + (n-1)L_b$$

Notice that these bounds are extremely simple to calculate, requiring no iteration.

12.7   Performance bound hierarchies

Although balanced job bounds are more accurate than asymptotic bounds, both suffer from an 'all or nothing' problem. Either the bounds give acceptable accuracy, or they do not. The accuracy of the bounds is not open to improvement. Performance bound hierarchies allow arbitrary levels of accuracy to be obtained, depending on the amount of calculation undertaken. The technique applies to single class separable queueing networks with only fixed rate server or infinite server nodes. For computational purposes, the infinite server nodes can be coalesced into a single infinite server node. This infinite server node is labelled as node 0, and the other fixed rate servers as nodes 1 to M. The total load at node 0, and hence the delay at that node, will be denoted by L_0.

Applying MVA, the delay at node k when the population is n can be expressed in terms of network performance when the population is $n-1$.

$$R_k(n) = L_k \left[1 + (n-1)\frac{R_k(n-1)}{L_0 + R(n-1)} \right]$$

Approximate MVA uses this relationship to find the delay through the network at population n, estimates the delays at population $n-1$, and iterates its approximations until the relationship is satisfied. Performance bound hierarchies are based on exact MVA. The ith level performance bound on delay at node k when the network population is n is calculated using the $i-1$st set of bound s for the network with population $n-1$. Denoting the ith level bounds with a superscript (i),

$$R_k^{(i)}(n) = L_k \left[1 + (n-1)\frac{R_k^{(i-1)}(n-1)}{L_0 + R^{(i-1)}(n-1)} \right]$$

and:

$$R^{(i)}(n) = \sum_{k=1}^{M} R_k^{(i)} \tag{12.9}$$

In order to calculate level i bounds, bounds at level 0 must be chosen for the network with population $n-i$. The choice of $R_k^{(0)}(n-i)$ determines whether the bounds on delays are optimistic, i.e. lower bounds, pessimistic, i.e. upper bounds, or just approximations.

12.7.1   Optimistic hierarchy

An initial choice of $R_k^{(0)}(n)$ using asymptotic bound analysis results in an optimistic set of bounds. That is, the asymptotic lower bound on total system delay is apportioned equally among all nodes:

$$R_k^{(0)}(n) = \frac{1}{M} \max[nL_b - Z, R(1)] \tag{12.10}$$

Clearly, we have:

$$R^{(0)}(n) = \max[nL_b - Z, R(1)]$$

which is the asymptotic lower bound on system response time. The estimate of delay at each node, (12.10), is not uniformly optimistic or pessimistic. For heavily loaded nodes it is optimistic, for lightly loaded nodes, pessimistic. It is possible to prove that $R(n) \geq R^{(i)}(n)$, but *not* that $R^{(i+1)}(n) \geq R^{(i)}(n)$. In order for that condition to hold, if $R^{(i+1)}(n)$ is replaced by $R^{(0)}(n)$ if the asymptotic bound is larger.

12.7.2   Pessimistic hierarchy

A pessimistic hierarchy results when the delay is assumed to occur entirely at the bottleneck node, that is:

$$R_b^{(0)}(n) = nR(1)$$
$$R_k^{(0)}(n) = 0, \quad k \neq b$$

This is pessimistic (extremely!) at the bottleneck node, but optimistic elsewhere. This initialisation gives:

$$R^{(0)}(n) = nR(1)$$

Again it can be shown that $R^{(i)}(n) \geq R(n)$, but in the pessimistic case $R^{(i+1)}(n) \leq R^{(i)}(n)$, so the sequence of upper bounds on the response time is strictly decreasing.

12.8   Further reading

The original approximate MVA algorithm was due to Bard[3]. Linearizer was developed by Neuse and Chandy[9]. The techniques described here have been extended to admit load dependent servers by Zahorjan and Lazowska[25] PAM algorithms were first described by Hsieh and Lam[13]. Agrawal's book[1] is an excellent description and comparison of a number of approximation techniques.

Flow equivalent servers were investigated by Chandy, Herzog, and Woo[8]. They prove the correctness of Norton's theorem for BCMP networks, and in a later paper use it to investigate solution of general queueing networks[7]. Chandy and Sauer also provided a survey of the state of the art in approximate solution techniques in[10]. Balsamo and Iazeolla have proved some extensions to Norton's theorem[2].

Attempts to approximate priority queueing networks in a BCMP framework have been underway ever since the original BCMP paper was published[4]. The reduced work-rate approximation was used by Reiser[22]. Shadow approximations were investigated by Sevcik[23], although he only applied them to networks with a single priority server. The MVA priority approximation is the work of Bryant et al[6]. The later corrections are due to Kaufman[14] and Bondi and Chuang[5].

Asymptotic bound analysis is described in Kleinrock's book[15, Chapter 4]. Balanced job bounds were developed by Zahorjan et al[26]. The system of bound hierarchies is the work of Eager and Sevcik[11]. Similar bounding analysis has also been done by Suri[24]. Sevcik and Eager extended their analysis to multiclass networks in [12].

Separable networks with large numbers of customers can be efficiently analysed using asymptotic expansions of integral representations of the partition function. This approach was first investigated by McKenna and Mitra[19] and has been extensively studied by them[17, 18, 20, 21]. Although an approximation technique, lower and upper bounds are provided to the performance measures. Recently Knessl and Tier have developed a different sort of asymptotic expansion technique[16].

12.9   Bibliography

[1] S.C. Agrawal. *Metamodeling: A Study of Approximations in Queueing Models*. MIT Press, Cambridge, Massachusetts, 1985.

[2] S. Balsamo and G. Iazeolla. An extension of Norton's theorem for queueing networks. *IEEE Transactions on Software Engineering*, SE-8(4):298–305, 1982.

[3] Y. Bard. Some extensions to multiclass queueing network analysis. In M. Arato, A. Butrimenko, and E. Gelenbe, editors, *Performance of Computer Systems*, pages 51–61. North-Holland, Amsterdam, 1979.

[4] F. Baskett, K.M. Chandy, R.R. Muntz, and F. Palacios-Gomez. Open, closed and mixed networks of queues with different classes of customers. *Journal of the ACM*, 22(2):248–260, April 1975.

[5] A.B. Bondi and Y.-M. Chuang. A new MVA-based approximation for closed queueing networks with a preemptive priority server. *Performance Evaluation*, 8(3):195–221, 1988.

[6] R.M. Bryant, A.E. Krzesinski, M.S. Lakshmi, and K.M. Chandy. The MVA priority approximation. *ACM Transactions on Computer Systems*, 2(4):335–359, November 1984.

[7]   K.M. Chandy, U. Herzog, and L.S. Woo. Approximate analysis of general queueing networks. *IBM Journal of Research and Development*, 19(1):43–49, January 1975.

[8]   K.M. Chandy, U. Herzog, and L.S. Woo. Parametric analysis of queueing network models. *IBM Journal of Research and Development*, 19(1):36–42, January 1975.

[9]   K.M. Chandy and D. Neuse. Linearizer: A heuristic algorithm for queueing network models of computing systems. *Communications of the ACM*, 25(2):126–134, February 1982.

[10] K.M. Chandy and C.H. Sauer. Approximate methods for analyzing queueing network models of computer systems. *ACM Computing Surveys*, 10(3):281–317, September 1978.

[11] D.L. Eager and K.C. Sevcik. Performance bound hierarchies for queueing networks. *ACM Transactions on Computer Systems*, 1(2):99–116, May 1983.

[12] D.L. Eager and K.C. Sevcik. Bound hierarchies for multiple-class queueing networks. *Journal of the ACM*, 33(1):179–206, January 1986.

[13] C.-T. Hsieh and S.S. Lam. PAM — a noniterative approximate solution method for closed multichain queueing networks. *Performance Evaluation*, 9(2):119–133, 1989.

[14] J.S. Kaufman. Approximation methods for networks of queues with priorities. *Performance Evaluation*, 4(3):183–198, 1984.

[15] L. Kleinrock. *Queueing Systems Volume 2: Computer Applications*. Wiley, New York, NY, 1976.

[16] C. Knessl and C. Tier. Asymptotic expansions for large closed queueing networks. *Journal of the ACM*, 37(1):144–174, January 1990.

[17] J. McKenna and D. Mitra. Integral representations and asymptotic expansions for closed Markovian queueing networks: Normal usage. *Bell System Technical Journal*, 61(5):661–683, 1982.

[18] J. McKenna and D. Mitra. Asymptotic expansions and integral representations of queue lengths in closed Markovian networks. *Journal of the ACM*, 31(2):346–360, April 1984.

[19] J. McKenna, D. Mitra, and K.G. Ramakrishan. A class of closed Markovian queueing networks: Integral representations, asymptotic expansions and generalizations. *Bell System Technical Journal*, 60(5):599–641, 1981.

[20] D. Mitra and J. McKenna. Some results on asymptotic expansions for closed Markovian networks with state dependent service rates. In E. Gelenbe, editor, *Performance '84*, pages 377–392. North Holland, Amsterdam, 1984.

[21] D. Mitra and J. McKenna. Asymptotic expansions for closed Markovian networks with state-dependent service rates. *Journal of the ACM*, 33(3):568–592, July 1986.

[22] M. Reiser. A queueing network analysis of computer networks with window flow control. *IEEE Transactions on Communications*, COM-27(8):1199–1209, August 1979.

[23]   K.C. Sevcik. Priority scheduling disciplines in queueing network models of computer systems. In B. Gilchrist, editor, *Information Processing 77*, pages 565–570. North Holland, Amsterdam, 1977.

[24]   R. Suri. Generalized quick bounds for performance of queueing networks. *Computer Performance*, 5(2):116–120, 1984.

[25]   J.L. Zahorjan and E.D. Lazowska. Incorporating load dependent servers in approximate mean value analysis. *Performance Evaluation Review*, 12(3):52–62, 1984.

[26]   J.L. Zahorjan, K.C. Sevcik, D.L. Eager, and B.I. Galler. Balanced job bound analysis of queueing networks. *Communications of the ACM*, 25(2):134–141, February 1982.

Chapter 13

Numerical solution of queueing models

The emphasis of the previous chapters has been on finding closed form expressions for the various performance measures in terms of the parameters of the model being studied. Although closed forms are attractive to the analyst and provide insight into the behaviour of the system, ultimately the numerical evaluation of the performance measures is the aim of analysing systems. If the model can be set up in such a way that it can be solved numerically from the start, it may be possible to find the performance of systems for which no closed form solutions are available.

BCMP networks are a good example. Although they have been applied successfully to many problems, they cannot deal with some very important types of queueing network behaviour. Perhaps the most important restriction is that all classes of job must have the same (exponential) service distribution at FIFO stations. In reality, different classes of jobs *will* have different service requirements. Another important aspect of real life networks which is not within the BCMP framework is stations which give priority to one class of jobs or another. The analysis of BCMP networks is based on Markov processes. The underlying Markov process has a very complex state space, which is constrained in order that a product form solution results. If the constraints necessary for product form solution are not satisfied, one still has a (multidimensional) Markov process.

In this chapter we shall examine some approaches to numerical solution of queueing models. Since all stochastic processes can be approximated by Markov processes, with appropriate mapping of state spaces, we deal mostly with numerical solution of Markov processes and Markov chains. Observe that in the rest of this book, row vectors have been used throughout, without comment. In practice, all numerical subroutine libraries, and most literature references in numerical linear algebra deal with column vectors, and solve equations in the form $A\mathbf{x} = \mathbf{b}$. To solve problems involving row vectors and post-multiplication by matrices, the whole equation can be transposed. Effectively, the same equation is solved, except using the transpose of the matrix A. To minimise confusion when consulting linear algebra texts, we shall adopt the convention that vectors are column vectors and that row vectors are denoted by a superscript T, for this chapter only.

The definition of the stationary distribution of a Markov process can be used as a starting point.

$$\pi^T Q = \mathbf{0}^T \tag{13.1}$$

These balance equations are linear in the steady state probabilities and can be solved by a number of techniques.

The problem can also be presented in terms of eigenvalues and eigenvectors. Consider the equations:

$$\pi^T (I + Q\tau) = \pi^T \tag{13.2}$$

π can be considered as the left eigenvector corresponding to the unit eigenvalue of $P = I + Q\tau$. If τ is chosen such that $\tau < \max_i |q_{ii}|$, then P will be a stochastic matrix, and have all row sums equal to 1. Trivially, the column vector, $\mathbf{e}$, which has all elements equal to 1, is a right eigenvector of P, so 1 is an eigenvalue. Consideration of Gerschgorin's theorem shows that 1 is the largest modulus eigenvalue. P can be considered as the transition matrix of a Markov chain which makes transitions at intervals τ.

These two expressions give rise to different approaches to calculating π. Equation 13.2 shows that π is the left eigenvector of P corresponding to the unit eigenvalue. Equation 13.1 gives a homogeneous set of simultaneous linear equations for π. In both cases, the normalisation condition must be added to the equations to ensure that the solution vector is indeed a probability distribution. In Markov chains that arise in modelling queueing systems, it is usually the case that the transition matrix, or the rate matrix, is sparse, that is transitions from any state are only made to a few other states, so that the matrix will have a large number of zero entries. The number of states, n, is usually so large as to make storing the complete matrix impractical, so only the non-zero elements are stored. This sparsity has important implications for our choice of algorithm for the solution of the equations. For simplicity, we will also assume that the Markov chain is ergodic. Transient states are of little interest when the steady state distribution is sought, and can usually be eliminated. Similarly, if a process has more than one communicating class of states, each class can be analysed separately.

13.1 Homogenous equations

In this section we examine methods of solution of the homogenous set of equations (13.1). The matrix Q is singular, since the sum of all rows is 0. The following assumes that $\mathbf{e}_j$ is the column vector with all entries equal to 0, except for the jth which has the value 1, and that $\mathbf{e}$ is a column vector with all entries equal to 1. The equations can be solved directly by replacing one of them by the normalisation condition $\sum_{i=1}^{n} \pi_k = 1$ and solving by one of the standard methods. This is equivalent to replacing one column of Q, the kth say, with a column of 1s, and then replacing the zero vector on the right hand side of the equations by the unit vector, $\mathbf{e}_k$.

In practice, the transition probability matrix, P, is used. It has the advantage that all the elements are positive and are also less than 1 in magnitude. This is also a singular matrix, so a single equation must be replaced by the normalisation condition. Now P is a stochastic matrix, that is:

$$P\mathbf{e} = \mathbf{e}$$

The steady state vector that we are seeking satisfies:

$$\boldsymbol{\pi}^T P = \boldsymbol{\pi}^T \tag{13.3}$$

and the normalisation condition:

$$\boldsymbol{\pi}^T \cdot \mathbf{e} = 1 \tag{13.4}$$

13.1.1   Wachter's algorithm

This algorithm combines equations (13.3) and (13.4) to arrive at a formulation of the problem which is non-singular without the arbitrary choice of an equation to discard. Any vector $\mathbf{p}$, which satisfies equations (13.3) and (13.4) also satisfies:

$$\mathbf{p}^T[I - P + \mathbf{e}\mathbf{x}^T] = \mathbf{x}^T, \quad \forall \mathbf{x} \tag{13.5}$$

Notice that $\mathbf{e}\mathbf{x}^T$ is a matrix with all rows identical. Now this is non singular if $|I - P + \mathbf{e}\mathbf{x}^T| \neq 0$. Since $I - P$ is singular, $|I - P| = 0$. If the eigenvalues of $I - P$ are $\lambda_1, \lambda_2, \ldots, \lambda_n$, then one of them λ_1 say, equals 0. This is because the eigenvalues of the matrix are the roots of the characteristic polynomial, $|I - P - \alpha I| = (\alpha - \lambda_1) \cdots (\alpha - \lambda_n)$. Evaluated at $\alpha = 0$ this is the determinant of $I - P$, which is 0. Also $(I - P)\mathbf{e}^T = 0$. Now consider the characteristic polynomial of $I - P + \mathbf{e}\mathbf{x}^T$:

$$|I - P + \mathbf{e}\mathbf{x}^T - \alpha I| = |(I - P - \alpha I) \cdot (I + (I - P - \alpha I)^{-1}\mathbf{e}\mathbf{x}^T)|$$

The determinant $|I + \mathbf{y}\mathbf{x}^T| = 1 + \mathbf{x}^T\mathbf{y}$, for all vectors $\mathbf{x}, \mathbf{y}$. It follows that:

$$|I - P + \mathbf{e}\mathbf{x}^T - \alpha I| = |(I - P - \alpha I)| \cdot (1 + \mathbf{x}^T(I - P - \alpha I)^{-1}\mathbf{e})$$

Now $(I - P)\mathbf{e} = 0$, so $(I - P - \alpha I)\mathbf{e} = -\alpha\mathbf{e}$. It follows that $(I - P - \alpha I)^{-1}\mathbf{e} = -(1/\alpha) \cdot \mathbf{e}$, so long as α is not an eigenvalue of $I - P$. Hence:

$$|I - P + \mathbf{e}\mathbf{x}^T - \alpha I| = (1 + \mathbf{x}^T\mathbf{e} \cdot (1/\alpha)) \cdot \prod_{i=1}^{n}(\alpha - \lambda_i)$$

but $\lambda_1 = 0$, so the characteristic polynomial is:

$$|I - P + \mathbf{e}\mathbf{x}^T - \alpha I| = (\mathbf{x}^T\mathbf{e} - \alpha) \prod_{i=2}^{n}(\alpha - \lambda_2)$$

Evaluating this for $\alpha = 0$ shows that $I - P + \mathbf{e}\mathbf{x}^T$ is non-singular provided that $I - P$ has only a single zero eigenvector, and that $\mathbf{x}^T \cdot \mathbf{e} \neq 0$. Popular choices for $\mathbf{x}^T$ are $\mathbf{e}_j^T$ and $\mathbf{e}_j^T P$. If $\mathbf{e}_j^T P$ is chosen, then the jth row of $I - P + \mathbf{e}\mathbf{x}^T$ is $\mathbf{e}_j^T$. Use of Gaussian elimination with partial pivoting is recommended to solve (13.5).

13.1.2   Plemmon's algorithm

Although the instantaneous transition rate matrix, Q, and the transition matrix, P, are singular, it is possible to solve the balance equations (13.1). Instead of using Q directly, $I - P$ is used, because this formulation has effectively scaled the matrix elements. All diagonal elements of $I - P$ are positive, and the off-diagonal elements are all non-positive. The row sums are all identically 0. Equation (13.1) can be written as:

$$(I - P^T)\mathbf{p} = \mathbf{0}$$

Although the matrix is singular, a consistent solution to the equations can be found by Gaussian elimination on the matrix $I - P^T$. The matrix has a factorisation LU, where L is a lower triangular matrix with unit diagonal elements, and U is an upper triangular matrix with $u_{nn} = 0$. This factorisation is found using the normal Gaussian elimination algorithm. Since the diagonal elements are the largest in each column of $I - P^T$, pivoting is not necessary. Assuming the process is ergodic, the final elimination step should convert the nn element to 0. The magnitude of the non-zero value that is actually be found is an indication of the accuracy of the elimination process. Replacing u_{nn} by 1, we can solve the equations $U\mathbf{p} = \mathbf{e}_n$, by back substitution, and then normalise the vector $\mathbf{p}$ to find a probability distribution. Backward error analysis of the algorithm shows that the computed solution is the exact solution to the system $(I - P^T + E)\mathbf{p} = \mathbf{0}$. The elements of E satisfy $|e_{ij}| \leq \varepsilon i(3.02 + 1.01n)$, where ε is the smallest representable number such that $1 + \varepsilon > 1$.

13.1.3   Grassmann's algorithm

One of the main sources of error in numerical calculation is *cancellation error*, caused by the subtraction of nearly equal numbers. Grassmann's algorithm is essentially a rearrangement of Gaussian elimination to avoid all subtractions. The derivation of the algorithm depends on the regenerative properties of Markov chains.

Consider a Markov chain $X = \{X_n; n = 0, 1, \ldots\}$. The state space is assumed to be the non-negative integers $\{0, 1, 2, \ldots\}$. Assuming that the chain is ergodic, a steady state will be reached, which is independent of the starting state of the chain. Denote by $P_i(X_n = j) = \Pr(X_n = j|X_0 = i)$, and by $E_i[X_n] = \sum j P_i(X_n = j)$. The subscript is used to indicate the starting state. Because the chain is ergodic, we know that, for all i:

$$P_i(X_n = j) \to \pi_j \text{ as } n \to \infty$$

Consider a set of states $D = \{0, 1, 2, \ldots, d - 1\}$. Define a *sojourn from state* k as a first passage from state k to some state in the set D, that is a sample path which includes no states in the set D, except the final state in the sample path. Define $v_{ki}^{(d)}$ as the expected number of visits to state i in a sojourn from state k. If both i and k are in the set D then $v_{ki}^{(d)} = \delta(k, i)$, since the starting state is always in the sojourn. Usually $i \notin D$ and $k \in D$.

If the steady state distribution is π, it can be shown to satisfy:

$$\pi_i = \sum_{k=0}^{d-1} v_{ki}^{(d)} \pi_k \quad i = 0, 1, 2, \ldots$$

This is trivially so for $i < d$. When $i \geq d$, renewal theory arguments are used. We do not prove them here. Similarly,

$$v_{kj}^{(d)} = V_{kj}^{(d+1)} + v_{dj}^{(d+1)} v_{kd}^{(d)}, \quad k \neq d$$

$$v_{kj}^{(d)} = v_{kj}^{(j)} + \sum_{m=d}^{j-1} v_{km}^{(d)} v_{mj}^{(j)}, \quad j \geq d$$

Assuming a finite state space, $\{0, 1, \ldots, n\}$, the algorithm works on the matrix $A = P - I$. The nth equation for π_n, expressing it in terms of π_i for $i = 0, 1, \ldots, n - 1$. This symbolic expression is then used to eliminate π_n from all the other equations. The algorithm continues by eliminating π_{n-1} in a similar manner. At each stage, π_k is expressed in terms of $\pi_{k-1}, \pi_{k-2}, \ldots, \pi_0$. This elimination continues until only π_0 remains, and the normalisation condition is then used to find its exact value. Denoting $a_{ij}^{(m)}$ for the ij element of the matrix A, *before* the elimination of π_m, then the elimination is:

$$a_{im}^{(m-1)} = -a_{im}^{(m)}/a_{mm}^{(m)}, \quad 0 \leq i < m$$
$$a_{ij}^{(m-1)} = a_{ij}^{(m)} + a_{mj}^{(m)} a_{im}^{(m-1)}, \quad 0 \leq i < n, 0 \leq j < n$$
$$a_{ij}^{(m-1)} = a_{ij}^{(m)} \quad i \geq m \text{ or } j > m$$

Although this procedure appears to require subtractions, it can be shown that $a_{mm}^{(m)} = -\sum_{j=0}^{m-1} a_{ij}^{(m)}$. After making this substitution, and taking account of the fact that the diagonal elements are negative, one can express the procedure as an algorithm.

The algorithm is:

1. Initialise the array a to the transition matrix.
2. For $m = n, n - 1, \ldots, 1$ do

   (a) $S := \sum_{j=0}^{m-1} a_{mj}$

   (b) $a_{im} := a_{im}/S \quad \forall i < m.$

   (c) $a_{ij} := a_{ij} + a_{im} a_{mj} \quad \forall i, j < m.$

3. $TOT := 1$ and $r_0 := 1$.
4. For $j = 1, 2, \ldots, n$ do

   (a) $r_j := a_{0j} + \sum_{k=0}^{j-1} r_k a_{kj}$

   (b) $TOT := TOT + r_j$

5. $\pi_i := r_i/TOT \quad i = 0, 1, \ldots n$

This algorithm needs no subtraction, and has an operation count approximately twice that of standard Gaussian elimination.

13.1.4   Iterative techniques

The large state space makes direct solution methods unattractive because they tend to cause matrix fill-in, that is, zero elements which were not stored, become non-zero due to the effects of the Gaussian (or other) elimination. Using iterative methods to solve the equations avoids this problem, since the matrix is not altered. There is an extra bonus too, if we already know a good estimate to the steady state solution, then we can use that estimate to start our iterations and expect speedy convergence.

The main iterative techniques for solving linear equations are Jacobi iteration and Gauss-Seidel iteration. Either technique can be applied to equation (13.1), or to the equivalent using $I - P$ instead of Q. Using Q, the steady state probability distribution vector $\boldsymbol{\pi}$ satisfies:

$$q_{ii}\pi_i = \sum_{j=1}^{i-1} q_{ji}\pi_j + \sum_{j=i+1}^{n} q_{ji}\pi_j \tag{13.6}$$

Both iterative procedures involve choosing an initial probability distribution $\boldsymbol{\pi}(0)$, and calculating a new approximation $\boldsymbol{\pi}(1)$ based on solving (13.6). This process continues until the new estimate $\boldsymbol{\pi}(k + 1)$ differs from the old estimate, $\boldsymbol{\pi}(k)$, by less than some error bound. Jacobi iteration consists of applying equation (13.6) in parallel to $\boldsymbol{\pi}(k)$ to calculate $\boldsymbol{\pi}(k + 1)$:

$$\pi_i(k + 1) = \frac{1}{q_{ii}} \left(\sum_{j \neq i} q_{ji}\pi_j(k) \right)$$

This will require two vectors to hold the current and the old estimates for π.

Gauss-Seidel iteration starts from the same basic relationship, but always uses the most up to date estimate of π_i in its calculations. Assuming that the new estimates are calculated in the order 1, 2, 3, ..., the iterative equation is

$$\pi_i(k+1) = \frac{1}{q_{ii}} \left(\sum_{j=1}^{i-1} q_{ji}\pi_j(k+1) + \sum_{j=i+1}^{n} q_{ji}\pi_j(k) \right)$$

In both methods, the solution must be renormalised periodically.

Unfortunately, because the Q matrix is *not* diagonally dominant, convergence of either algorithm cannot be guaranteed. A *shifted Gauss-Seidel* algorithm is necessary to ensure convergence. Given an initial probability distribution $\pi(0)$ calculate $\mathbf{x}(0)$ using:

$$x_i(0) = \sum_{j=1}^{i} q_{ji}\pi_j(0)$$

Subsequent estimates are produced using:

$$x_i(k+1) = (1-\tau)x_i(k) + \tau \sum_{j=i+1}^{n} q_{ji}\pi_j(k)$$

$$\pi_i(k+1) = \frac{1}{q_{ii}} \left(x_i(k+1) - \sum_{j=1}^{i-1} q_{ji}\pi_j(k+1) \right)$$

The parameter τ can be chosen between 0 and 1.

13.2   Eigenvector solutions

This section examines methods of numerical solution based on the interpretation of π as a left eigenvector of P, (13.2). It is not usual to use direct methods to find the eigenvector, such as the QR algorithm, because they offer no complexity advantage over Gaussian elimination. Iterative methods that preserve the sparsity pattern of the transition matrix are preferred. From the definition of the transition matrix, if the system is in state i at time 0, then the probability that it will be in state j after one transition is given by p_{ij}. In general, if $\mathbf{p}$ is the probability distribution at some time, then $\mathbf{p}P$ is the distribution after one more transition. A straightforward algorithm for calculating π is to initialise a vector $\mathbf{p}$ to some probability distribution, and repeatedly post multiply it by the matrix P, until it is not changing. This is in some sense a simulation of the Markov chain, where the changing probability distribution is simulated. Since we assumed the ergodicity and regularity of the chain, convergence to a fixed point vector is guaranteed and we know that the steady state probability distribution, which is a fixed point, is unique. This technique for solving Markov chains is precisely the power method for finding the dominant eigenvalue and eigenvector of a matrix. This is a well known method, whose rate of convergence is proportional to the ratio of the dominant eigenvalue, 1 in this case, to the next largest eigenvalue in modulus.

Many models set up to solve computer configuration problems exhibit very slow convergence when the power method is used to find the steady state distribution. The cause is often the presence of sets of states which have a high rate of transition among themselves, and relatively rare transitions to other sets of states. For example, if a model incorporating the paging demands of a number batch processed jobs is set up, the states which correspond to a particular set of jobs being active, but with the allocation of memory to jobs being different between the different states will make transitions between states in the same set very frequently, but only make transitions to another set when a job terminates, or a new job starts. Investigation of the eigenvalues of the transition matrix will show that there are as many eigenvalues close to 1 as there are sets of states which frequently interact. Lopsided simultaneous iteration is a technique which simultaneously approximates the m dominant eigenvalues and eigenvectors of a matrix. Choosing the value of m to be larger than the number of strongly interacting subsystems will enable fast convergence to the dominant eigenvector (which is of interest) as well as to the sub-dominant ones (which are of no interest).

We use without comment the fact that the eigenvectors of a matrix are mutually orthogonal, since the conditions under which this is not so rarely arise in modelling applications.

13.2.1   Power method

As remarked above, the power method consists of repeatedly post multiplying the current estimate for the probability distribution by the transition matrix. Assuming the ergodicity of the Markov chain, the sequence of vectors given by:

$$\mathbf{p}(i+1) = \mathbf{p}(i)P$$

converges to the stationary probability distribution of P, $\boldsymbol{\pi}$. This can be seen by considering the eigenvalues and eigenvectors of P. Without loss of generality, assume that the eigenvalues of P are $\lambda_1, \lambda_2, \ldots, \lambda_n$, in order of decreasing magnitude. That is, $1 = \lambda_1 > \lambda_2 \geq \lambda_3 \cdots \geq \lambda_n$. The corresponding (left) eigenvectors are $\mathbf{v}_1, \mathbf{v}_2, \ldots, \mathbf{v}_n$. Of course, $\mathbf{v}_1 = \boldsymbol{\pi}$. Express $\mathbf{p}(0)$ as a linear combination of the left eigenvectors of P:

$$\mathbf{p}(0) = \alpha_1 \mathbf{v}_1 + \alpha_2 \mathbf{v}_2 + \cdots + \alpha_n \mathbf{v}_n$$

Hence, $\mathbf{p}(1) = \mathbf{p}(0)P$ satisfies

$$\mathbf{p}(1) = \alpha_1 \lambda_1 \mathbf{v}_1 + \alpha_2 \lambda_2 \mathbf{v}_2 + \cdots + \alpha_n \lambda_n \mathbf{v}_n$$

Continuing the same process,

$$\mathbf{p}(i) = \alpha_1 \lambda_1^i \mathbf{v}_1 + \alpha_2 \lambda_2^i \mathbf{v}_2 + \cdots + \alpha_n \lambda_n^i \mathbf{v}_n$$

As $i \to \infty$, $\lambda_j^i \to 0$ for $j = 2, 3, \ldots, n$ since $\lambda_j < 1$. $\lambda_1^i = 1$ for all i. Thus:

$$\mathbf{p}(i) \to \alpha_1 \mathbf{v}_1 \quad \text{as } i \to \infty$$

so that a simple renormalisation of the vector will produce the steady state probability distribution. The rate of convergence will be governed by the rate at which the powers of the non-unit eigenvalues approach zero. If the largest in magnitude of them, λ_2, is close to one, then convergence will be slow.

13.2.2   Simultaneous iteration

Simultaneous iteration attempts to improve the convergence of the eigenvector estimates by *simultaneously* approximating the m dominant eigenvectors of a matrix. Convergence will be governed by the ratio of the dominant eigenvalue, 1, in this case, to the eigenvalue which is not being approximated, λ_{m+1}. Traditional simultaneous iteration uses both left and right eigenvectors. We present an algorithm that approximates only one set.

For convenience, we shall evaluate the *right* eigenvectors. The steady state probability distribution is the dominant *left* eigenvector of the transition probability matrix, so to ensure that the steady state distribution is found, the algorithm is applied to P^T, the transposed matrix. A set of m initial trial vectors is chosen. If a solution to a related model is known, then its m dominant eigenvectors can be used. If starting from scratch, unit vectors can be used. Each trial vector is a linear combination of eigenvectors, so the set of trial vectors, arranged as an $n \times m$ matrix, can be expressed as a matrix product of the eigenvectors and a coefficient matrix. Writing U for the $n \times m$ matrix of trial vectors, E for the $n \times n$ matrix with each column consisting of an eigenvector, and C for the $n \times m$ matrix of coefficients:

$$U = E \cdot C$$

Assuming that the columns of E are ordered according to the magnitudes of the corresponding eigenvalue we write the eigenvalues as a diagonal matrix, Λ, with the eigenvalues in descending order of magnitude. If we partition the eigenvalues into the set $\mathcal{A}$, containing the m dominant eigenvalues, and $\mathcal{B}$ containing the others, we can also partition E into an $n \times m$ matrix $E_\mathcal{A}$, and an $n \times (n - m)$ matrix $E_\mathcal{B}$. The matrix of coefficients, C, is partitioned into $C_\mathcal{A}$ and $C_\mathcal{B}$ which are $m \times m$ and $(n - m) \times m$, respectively. The trial vectors can be represented as:

$$U = E_A C_A + E_B C_B$$

After multiplication by the (transposed) transition matrix,

$$V = P^T U = E_A \Lambda_A C_A + E_B \Lambda_B C_B$$

and the coefficients of E_B in V have been reduced more than the coefficients of E_A. Iterating using this form would only produce m vectors all converging on the steady state distribution at constant rate. The initial trial vectors would determine which converged first. Lopsided iteration (simultaneous iteration for one set of eigenvectors only) then calculates the interaction between the different eigenvectors, and uses the information to separate the different eigenvectors. Two $m \times m$ matrices are formed:

$$G = U^T \cdot U \quad \text{and} \quad H = U^T \cdot V$$

and the $m \times m$ *interaction* matrix B is calculated by solving:

$$GB = H$$

B relates the successive estimates of the trial vectors. It contains the essential information of P^T in some sense. The next step is to find the full set of (right) eigenvectors of B. A matrix W of these is calculated in the same way as the matrix of trial vectors. The first column of W is the dominant eigenvector of B, the second column, the sub-dominant eigenvector, and so on. A new estimate for the dominant eigenvectors of P^T is found by renormalising the matrix U^*, which is given by:

$$U^* = V \cdot W$$

This new set of trial vectors is then used as the value for U. This iteration continues until the vectors have converged.

It is interesting to observe why this procedure works. The matrices that are used to calculate the interaction matrix B are formed as follows:

$$G = U^T \cdot U = C_A^T C_A + C_B^T C_B$$
$$H = U^T \cdot V = U^T \cdot P^T U = C_A^T \Lambda_A C_A + C_B^T \Lambda_B C_B$$

If one assumes that the coefficients C_B are small in comparison to C_A, then:

$$C_A^T C_A B \approx C_A^T \Lambda_A C_A$$

and assuming that the matrix C_A^T is non singular,

$$C_A B \approx \Lambda_A C_A$$

The eigenvalues of B are an approximation to the m dominant eigenvalues of P, and C_A^T are an approximation to the left eigenvectors of B. The matrix, W, of right eigenvectors of B is an approximation to C_A^{-1}, and:

$$U^* = V \cdot W = E_A \Lambda_A + E_B \Lambda_B C_B C_A^{-1}$$

is a closer approximation to the dominant eigenvectors. This argument holds so long as C_B is small. We now demonstrate that the iteration is convergent. The effect of a single iteration is to calculate U^* from U.

$$U = E_A C_A + E_B C_B$$
$$U^* = (E_A \Lambda_A C_A + E_B \Lambda_B C_B) \cdot W \cdot N$$

where W is the matrix of eigenvectors of the interaction matrix, and N represents the normalisation operations. After k iterations, we have:

$$U^{*(k)} = (E_A \Lambda_A^k C_A + E_B \Lambda_B^k C_B) \cdot Z_k$$

where Z_k is the product of the eigenvector matrices and normalisation matrices at each iteration:

$$Z_k = W_1 N_1 W_2 N_2 \cdots W_k N_k$$

where W_k represents the eigenvectors of the kth interaction matrix B. Hence after k iterations, the coefficients of the eigenvectors have been multiplied by the corresponding eigenvalue k times. Since the eigenvalues corresponding to the coefficients C_A are larger than those corresponding to C_B, the coefficients $\Lambda_B C_B$ will eventually be swamped by $\Lambda_A C_A$.

13.3   Decomposition methods

Many computer systems have changes of state occurring at several different rates. For example, in a terminal system, the number of active terminals changes relatively infrequently, maybe a few times per hour. The number of jobs actually requesting service at the CPU system changes more rapidly, often several times per minute. Within the computer system itself, the number of jobs waiting for the CPU, and waiting for the disks will change more rapidly still, typically in the order of ten to one hundred times per second. If there are features of the system which preclude a product form solution, then a numerical solution is the only generally applicable solution technique.

Attempting to model all these interactions at once will cause a number of difficulties. First, the state space will very large. This almost certainly means that an iterative technique will be needed to find the steady state distribution. Assume that the power method is used[†]. Suppose also that the initial vector is a unit vector corresponding to the system having T active terminals, and $J \leq T$ active jobs. The

---

[†]Using Gauss-Seidel iteration will encounter similar problems.

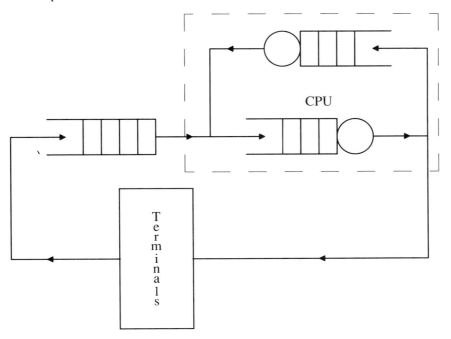

Figure 13.1 A queueing network

initial iterations will distribute the probability among the other states having T active terminals and J active jobs, but states with other numbers of terminals and/or jobs will still have a very low probability because the changes in those components of the state space happen at a much lower rate than changes in the internal disposition of jobs between CPU and disk queues. Essentially a steady state will be reached in the subsystem consisting of those states with T active terminals and J active jobs.

Consider the queueing network shown in Figure 13.1. There are N terminals, each of which goes through a 'think'-'compute' cycle. After thinking for a time which is exponentially distributed with parameter γ, the terminal submits a job to the CPU subsystem. Within the CPU subsystem, jobs join the CPU queue, and after an exponentially distributed service with parameter μ, return to the terminal and another think time with probability p. With probability $q = 1 - p$, a job leaving the CPU request I/O at the disk, which is exponentially distributed with parameter δ. We assume that $\gamma \ll \delta \approx \mu$, in magnitude and that $p \ll q$. The state of the system can be represented by a pair of integers, (i, j), with i representing the total number in the CPU subsystem, and j the number at the CPU. The number of jobs in think state at the terminals is given by $N - i$. The number of jobs at the disk is $i - j$. Clearly, i can have any value between 0 and N, inclusive, and j can take values between 0 and i, inclusive. If $N = 3$ the possible states of the system are (0,0), (1,0), (1,1), (2,0). (2,1), (2,2), (3,0), (3,1), (3,2), and (3,3). In general, there will be $(N + 1)(N + 2)/2$ states. As so far presented, the model is a BCMP network, and can be solved using the standard algorithms. A simple modification to the model will prevent this approach. If the multiprogramming level allowed is at most K, then a job attempting to leave a terminal will be blocked until the number of jobs at the CPU and I/O combined is less than K. In terms of the state representation we have adopted, it means that states with $i > K$ represent states in which K jobs are in the CPU subsystem, and $i - K$ are blocked waiting to enter it. States with $j > K$ are impossible. The number of jobs at the I/O device is $\max(i, K) - j$. It is clear that the transition matrix of the equivalent Markov chain will not only be sparse, but will have the following structure:

$$P = \begin{pmatrix} P_0 & U_0 & 0 & 0 & 0 & \cdots & 0 & 0 \\ D_1 & P_1 & U_1 & 0 & 0 & \cdots & 0 & 0 \\ 0 & D_2 & P_2 & U_2 & 0 & \cdots & 0 & 0 \\ 0 & 0 & D_3 & P_3 & U_3 & \cdots & 0 & 0 \\ \vdots & \vdots & \vdots & \vdots & \vdots & \ddots & \vdots & \vdots \\ 0 & \cdots & \cdots & \cdots & 0 & D_N & P_N \end{pmatrix}$$

where the P_i are $(i+1) \times (i+1)$ submatrices, when $i \le K$, and $(K+1) \times (K+1)$ submatrices when $i > K$. The submatrices U_i are $(i + 1) \times (i + 2)$ and the sub matrices D_i are $(i + 1) \times i$, when $i \le K$, and correspondingly altered for $i > K$. In fact, in this case, the P_i and D_i matrices will be identical for all $i \ge K$.

The magnitude of the entries in the P_i submatrices is much larger than the magnitude of the entries outside those submatrices.

A stochastic matrix, P, is said to be *nearly completely decomposable* if it can be expressed as:

$$P = P^* + \epsilon C, \quad \epsilon \ll 1 \tag{13.7}$$

with P^* a stochastic matrix of block diagonal form, and C a matrix with zero row sums. The submatrices of P^* which form the diagonal blocks are clearly stochastic themselves. The elements of C which fall within the diagonal blocks defined by the structure of P^* are all non-positive, and the elements outside those diagonal blocks are non-negative. Furthermore all elements of C are less than 1 in magnitude.

A first approximation to the steady state probability vector of P can be found by calculating the steady state probability vectors of the diagonal blocks of P^*. If ϵ is small, then these individual solution vectors will be a good approximation to the distribution of probability within the subsystems. The problem remains of how to normalise the vector so-formed. A naive approach of simply summing the elements will not do, since that will give equal weight to all the diagonal blocks. Clearly, some approximation to the slower rate dynamics of the system is needed. The transition rates *between* subsystems must be approximated. In our example, we wish to find the steady state probability associated with the transition matrix:

$$P' = \begin{pmatrix} p_0 & u_0 & 0 & 0 & 0 & \cdots & 0 & 0 \\ d_1 & p_1 & u_1 & 0 & 0 & \cdots & 0 & 0 \\ 0 & d_2 & p_2 & u_2 & 0 & \cdots & 0 & 0 \\ 0 & 0 & d_3 & p_3 & u_3 & \cdots & 0 & 0 \\ \vdots & \vdots & \vdots & \vdots & \vdots & \ddots & \vdots & \vdots \\ 0 & 0 & \cdots & \cdots & \cdots & 0 & d_N & p_N \end{pmatrix}$$

where p_i is the probability of *not* leaving the ith subsystem, u_i is the probability that the ith subsystem is left to the $i + 1$st subsystem, and d_i is the probability that it is left to enter the $i - 1$st subsystem. In this case, we can estimate these probabilities simply from our knowledge of the system. $u_i = i\gamma$, $d_i \approx p\mu_i$, and $p_i = 1 - u_i - d_i$. u_i is exact, only d_i is an approximation, because of the possibility that all the jobs at the CPU system are at the disk waiting for I/O.

A general approach to the problem of how to exploit near complete decomposability will now be outlined. Assume that an n state Markov chain with transition matrix P is nearly completely decomposable, such that relation (13.7) holds. The completely decomposable matrix P^* has N stochastic diagonal blocks, P_k^*, $k = 1, 2, \ldots N$. P_k^* is $m(k) \times m(k)$. Clearly, $n = \sum_{k=1}^{N} m(k)$. Let the set of states that fall within the diagonal block P_k^* be denoted by $\mathcal{I}_k = \{m(k-1)+1, \ldots, m(k)\}$, and call this the *macro state* k. Denote the steady state probability vector of P by $\boldsymbol{\pi}$, and that subsection of it corresponding to the kth macro state by $\boldsymbol{\pi}_k$. Let $\theta_k = \sum_{j \in \mathcal{I}_k} \pi_j$, that is, the steady state probability that the system is in macro state k. If the steady state probability vector corresponding to P_k^* is $\mathbf{p}_k$, then

$$\mathbf{p}_k \approx (\theta_k)^{-1} \boldsymbol{\pi}_k$$

Examining the behaviour of P again, we see that the macro state transition probabilities can be calculated. If the system was in state i previously then it will be in state j after one transition with probability p_{ij}. Hence the probability that it will be in macro state r is $\sum_{j \in \mathcal{I}_r} p_{ij}$. The joint probability of having been in block k and being in block r after a single transition is evaluated by summing over all states i in block k. The *conditional* probability is found by dividing by the probability of being in block k. Hence, the block interaction matrix P' is given by:

$$p'_{kr} = (\theta_k)^{-1} \sum_{i \in \mathcal{I}_k} \pi_i \sum_{j \in \mathcal{I}_r} p_{ij}$$

$$\approx \sum_{i \in \mathcal{I}_k} \mathbf{p}_{k,i} \sum_{j \in \mathcal{I}_r} p_{ij}$$

where $\mathbf{p}_{k,i}$ is the element of the $\mathbf{p}_k$ vector corresponding to state i. Applying this technique to our example above, the d_i entry in the macro state transition matrix would be replaced by *not* $p\mu$, but $p\mu \times \Pr(\text{CPU busy})$.

It can be shown that the aggregation proposed is exact in the case that the Markov chain is *lumpable*. A Markov chain is lumpable on a partition of its state space into non-intersecting subsets, $\mathcal{I}_k$, if its transition matrix, $P = (p_{ij})$ satisfies the following condition for all subsets $\mathcal{I}_k$:

$$\sum_{j \in \mathcal{I}_k} p_{ij} = \sum_{i \in \mathcal{I}_k} p_{mj} \quad \text{for all } i, m \in \mathcal{I}_r \text{ and for all subsets } \mathcal{I}_k \text{ and } \mathcal{I}_r$$

That is, choose a state, i say, and calculate the total transition rate out of that state to all the subsets $\mathcal{I}_k$. The chain is lumpable if these state to subset transition rates are identical for all states in the same subset as i, and this condition holds for all the subsets of which i could be a member.

13.4 Matrix-geometric methods

Many systems which have no easily derived closed form can be solved numerically by taking advantage of their inherent structure. If the state space can be described by a pair of integers, (i, j), with only i being unbounded, then it is possible that a *matrix-geometric* solution may exist. In order to simplify the discussion, a concrete example will be chosen first, and then the general case will be described.

Consider a pair of $M/M/1$ queues in tandem, but with a finite buffer space between them. When the buffer space is full, the first server cannot serve any jobs until a free space becomes available in the buffer. If the first server went on serving, and the departures from the first server which found the buffer full were then lost, the two queues can be analysed independently as an $M/M/1$ system and an $M/M/1/N$ system. If the buffer space between the two queues is unbounded, Burke's theorem applies, and they act as if they are independent $M/M/1$ systems. Let the arrival rate of jobs to the system be λ, and the service rates of the servers be μ_1 and μ_2, respectively. If the second server has buffer space for N jobs, including the one in service, then the system's state can be described by a pair of integers, (i, j), with i representing the number of jobs at the first server, including the job in service, and j representing the number of jobs at the second server. i is non-negative, and $0 \leq j \leq N$. Assuming that steady state is reached, we denote the steady state probability of state (i, j) by p_{ij}. The balance equations are easy to write down, recalling that when $j = N$, the first server is blocked, and does no work:

$$\lambda p_{00} = \mu_2 p_{01}$$
$$(\lambda + \mu_2) p_{0j} = \mu_1 p_{1,j-1} + \mu_2 p_{0,j+1}, \quad 0 < j < N$$

$$(\lambda + \mu_2)p_{0N} = \mu_1 p_{1,N-1}$$
$$(\lambda + \mu_1)p_{i0} = \lambda p_{i-1,0} + \mu_2 p_{i1}, \quad i > 0$$
$$(\lambda + \mu_2)p_{ij} = \lambda p_{i-1,j} + \mu_1 p_{i+1,j-1} + \mu_2 p_{i,j+1}, \quad 0 < j < N, \quad i > 0$$
$$(\lambda + \mu_2)p_{iN} = \lambda p_{i-1,N} + \mu_1 p_{i+1,N-1}, \quad i > 0$$

Solution of these equations could be undertaken using generating functions, in the same way as the $M/M/1$ queue with breakdowns and repairs in chapter 6. However, if we concentrate our attention on the unbounded component of the state space, i, we observe that changes in that component occur in steps of 1. This is similar to a birth-and-death process. If we order the states lexicographically, and define the vector $\mathbf{p}_i = (p_{i0}, p_{i1}, \ldots, p_{iN})$, that is, the steady state probabilities of those states which have first component i, then the balance equations can be written as:

$$B_0 \mathbf{p}_0 + A_2 \mathbf{p}_1 = \mathbf{0}$$
$$A_0 \mathbf{p}_{j-1} + A_1 \mathbf{p}_j + A_2 \mathbf{p}_{j+1} = \mathbf{0} \quad j > 0$$

This is known as a *quasi-birth-and-death process*. A_0 is the $N \times N$ diagonal matrix with all entries equal to λ, and B_0, A_1, and A_2 are $N \times N$ matrices, with the following structures:

$$B_0 = \begin{pmatrix} -\lambda & \mu_2 & \cdots & \cdots & & 0 \\ 0 & -(\lambda + \mu_2) & \mu_2 & \cdots & & 0 \\ 0 & \cdots & & \ddots & & \cdots & 0 \\ 0 & \cdots & & 0 & -(\lambda + \mu_2) & \mu_2 \\ 0 & \cdots & & \cdots & 0 & -(\lambda + \mu_2) \end{pmatrix}$$

$$A_1 = \begin{pmatrix} -(\lambda + \mu_1) & \mu_2 & \cdots & \cdots & & 0 \\ 0 & -(\lambda + \mu_1 + \mu_2) & \mu_2 & \cdots & & 0 \\ 0 & \cdots & & \ddots & & \cdots & 0 \\ 0 & \cdots & & \cdots & -(\lambda + \mu_1 + \mu_2) & \mu_2 \\ 0 & \cdots & & \cdots & 0 & -(\lambda + \mu_2) \end{pmatrix}$$

$$A_2 = \begin{pmatrix} 0 & 0 & \cdots & \cdots & 0 \\ \mu_1 & 0 & 0 & \cdots & 0 \\ 0 & \mu_1 & \ddots & \cdots & 0 \\ 0 & \cdots & \mu_1 & 0 & 0 \\ 0 & \cdots & 0 & \mu_1 & 0 \end{pmatrix}$$

This organisation of the balance equations does not appear to make any progress until the one step nature of the transitions in the i component is used. Let us assume that a steady state is reached, and consider the state (i, j). For $i \geq 1$, the system *must* have been in state $(i-1, k)$ for some k at some time in the past. Consider the most recent time that it was in one of these $i-1$ states. Since then there has been a single arrival, followed by a series of transitions to states (x, y) with $x \geq i$. Let $h_{jk}(t)$ be the probability of transition from state (i, k) to state (i, j) in time t, *without* visiting any state $(i-1, m)$. The key observation is that this is *independent* of i. Integrating over all possible t, we denote the resulting matrix by H. Considering the vector of state probabilities $\mathbf{p}_i$, the following relationship is clear:

$$\mathbf{p}_i = \lambda H \mathbf{p}_{i-1} = (\lambda H)^i \mathbf{p}_0$$

Letting $R = \lambda H$, the balance equations can now be rewritten as:

$$B_0 \mathbf{p}_0 + A_2 R \mathbf{p}_0 = \mathbf{0} \tag{13.8}$$
$$A_0 R^{j-1} \mathbf{p}_0 + A_1 R^j \mathbf{p}_0 + A_2 R^{j+1} \mathbf{p}_0 = \mathbf{0} \quad j > 0$$

$\mathbf{p}_0$ can be extracted as a common factor. In the second equation, R^{j-1} is also a common factor. Since $\mathbf{p}_0$ is not identically zero if a steady state exists, the following relationships must hold:

$$B_0 + A_2 R = \mathbf{0} \tag{13.9}$$
$$A_0 + A_1 R + A_2 R^2 = \mathbf{0} \tag{13.10}$$

At this stage, the value of R is not yet known. Calculation from its definition in terms of the matrix H is extremely difficult. It appears that equation (13.9) could be used, but A_2 is a singular matrix. If a solution to the (matrix) quadratic equation (13.10) can be found, then equation (13.8) can be used to find $\mathbf{p}_0$, with one value, p_{00} say, chosen arbitrarily. The normalisation condition is then used to find the true values. It can be shown that the Markov process will be positive recurrent if and only if the minimal non-negative solution, R, to equation (13.10) has all eigenvalues less than one in magnitude, *and* equation (13.9) has a solution $\mathbf{p}_0$ which satisfies:

$$\mathbf{e}^T (I - R)^{-1} \mathbf{p}_0 = 1$$

Solution of the matrix quadratic equation (13.10) can be achieved by successive substitution, starting with $R(0) = 0$ as a first approximation.

$$R(k+1) = (A_0 + (A_1 - \text{diag}(A_1))R(k) + A_2 R(k)^2)(-\text{diag}(A_1))^{-1}$$

This recurrence converges to the minimal non-negative solution. An alternative recurrence uses:

$$R(k+1) = -(A_0 + A_2 R(k)^2)A_1^{-1}$$

This has the disadvantage that explicit calculation of an inverse matrix is required, which is an expensive operation.

Once the matrix R and the empty state probability vector $\mathbf{p}_0$ have been found, it is straightforward to find the mean number in the system, and the mean number of jobs at server 1, and at server 2. The mean of the total number of jobs in the system, N, is calculated as follows:

$$N = \sum_{i=0}^{\infty} \sum_{j=0}^{n} (i+j) p_{ij}$$

$$= \sum_{i=0}^{\infty} \sum_{j=0}^{N} (i+j) R_j^i \mathbf{p}_0$$

where R_j^n is the jth row of the matrix R^n. Introduction of the (row) vectors $\boldsymbol{\theta}^T = (0, 1, 2, \ldots, N)$ and $i \cdot \mathbf{e}^T$ allows this to be written as:

$$N = \sum_{i=0}^{\infty} (\boldsymbol{\theta}^T + i \cdot \mathbf{e}^T) R^i \mathbf{p}_0$$

The knowledge that R has all eigenvalues less than one allows this to be written as:

$$N = (\boldsymbol{\theta}^T (I - R)^{-1} + R(I - R)^{-2}) \mathbf{p}_0$$

For quasi-birth-and-death processes, the stability condition can also be found from the transition matrix in block form. It is always the case that finite numbers of boundary conditions do not affect the stability of the system. It is the transition rates that apply to the infinite portion of the state space that determine stability or saturation. The finite generator $A = A_0 + A_1 + A_2$ is also the rate matrix of a finite state Markov process. Its steady state distribution vector $\boldsymbol{\pi}$ is (loosely) the long run marginal probability distribution of the bounded component of the state space. If the following condition is satisfied:

$$\mathbf{e}^T A_2 \boldsymbol{\pi} > \mathbf{e}^T A_0 \boldsymbol{\pi}$$

then the quasi-birth-and-death process is stable. The condition effectively insists that the average rate of increase in the unbounded component is less than the average rate of decrease in the same component.

When we move from a specific model to the general form of problems that can be solved by matrix geometric techniques, two structures of transitions are possible. In both cases, the boundary states may exhibit arbitrary transition structures, but beyond a certain state, the transition matrix (or rate matrix), exhibits one of two regular block structures. The first type is known as a *Markov chain of $M/G/1$ type*, which has a block structure of the form:

$$P = \begin{pmatrix} B_0 & B_1 & B_2 & B_3 & B_4 & \cdots \\ C_0 & A_1 & A_2 & A_3 & A_4 & \cdots \\ 0 & A_0 & A_1 & A_2 & A_3 & \cdots \\ 0 & 0 & A_0 & A_1 & A_2 & \cdots \\ \vdots & \vdots & \vdots & \vdots & \vdots & \vdots \end{pmatrix}$$

in which all the matrices A_i are square, of the same dimensions, B_0 is square, although possibly of different dimension to the A_i matrices, and the other matrices have their dimensions determined by the dimensions of B_0 and A_0. The equivalent of the matrix quadratic equation (13.10) is:

$$G = \sum_{i=0}^{\infty} A_i G^i$$

In general, the solution to this matrix equation is not unique. It is important that the minimal non-negative solution is chosen. Computation of the steady state probabilities and moments is by no means as direct as it is in the quasi-birth-and-death process example used above. The interested reader is referred to Neuts' excellent book.

The second type is a *Markov chain of the GI/M/1 type* in which the transition matrix has the following block structure, with all matrices square[‡]:

$$P = \begin{pmatrix} B_0 & A_0 & 0 & 0 & 0 & \cdots \\ B_1 & A_1 & A_0 & 0 & 0 & \cdots \\ B_2 & A_2 & A_1 & A_0 & 0 & \cdots \\ B_3 & A_3 & A_2 & A_1 & A_0 & \cdots \\ \vdots & \vdots & \vdots & \vdots & \vdots & \vdots \end{pmatrix}$$

In this case, the steady state probability vector has matrix geometric form, with:

$$\mathbf{p}_i^T = \mathbf{p}_0^T R^i$$

where R is the minimal non-negative solution of the matrix equation

$$R = \sum_{i=0}^{\infty} R^i A_i$$

The solution of this equation can be carried out in the same way as the matrix quadratic (13.10), by successive substitution, either directly, or after the extraction of the 'linear' term RA. Moments can often be extracted by direct consideration of the matrix geometric form, and the values of R and $\mathbf{p}_0$.

[‡]These matrices are actually the transposes of the ones given in our example.

13.5 Exercises

1. Consider an $M/M/1/N/K$ system. Find its steady state distribution, and mean queue lengths in symbolic form. Compare the accuracy and speed of solution with numerical techniques. Solve it numerically, using Wachter's method, Grassmann's method, and the power method. Investigate the effect of increasing N on all solution methods. If a library sparse matrix routine is available, compare its speed at solving this problem with the power method's speed.

2. Many queues do not operate in a stable environment. Arrival rates may depend on time of day for example. Investigate the $M/M/1$ queue in which the arrival and service rates are themselves random variables. The arrival and service parameters are controlled by an environment process with N states. When the environment process changes state, the rate of service for a job in service changes at the same instant. If the environment process remains in a particular state for an exponentially distributed length of time, show that a solution can be calculated by matrix-geometric techniques. Investigate the behaviour of the $M/M/1$ queue with a two state environment with $\lambda_1 = 10$, $\mu_1 = 8$, $\lambda_2 = 3$, and $\mu_2 = 6$, when the environment process stays in each state for an interval of mean length 1.

13.6 Further reading

The numerical solution of the steady state vector of Markov chains has been recognised as an interesting problem for a number of years, but recently has been a focus of much interest.

Rosenberg and Wallace[20] proposed a system which used Jacobi iteration to solve the balance equations numerically. The approach given here which automatically ensures satisfaction of the normalisation condition is due to Paige, Styan and Wachter[14]. The Plemmon's algorithm was well known, but a strict backward error analysis is given in [8]. Grassmann's algorithm is developed in [7]. Heyman[9] extended Harrod and Plemmon's comparisons to include Grassmann's algorithm, and presented numerical results.

Analysis of the eigenvalues and eigenvectors of the transition matrix is due to W.J. Stewart, who developed lopsided iteration[16] and compared it with a number of other techniques[17].

Near complete decomposability was first applied to computing systems by Courtois[3] and has been treated in great detail[4]. A more philosophical discussion appears in[5]. Examples of its application to queueing networks are due to Vantilborgh[18, 19], Balsamo and Iazeolla[1], and Cao and W.J. Stewart[2]. Error bounds are discussed by G.W. Stewart[15].

Matrix-geometric recurrences were observed by Evans[6], but their study has been mostly carried out by M.F. Neuts and his students. He is responsible for one book which treats the $G/M/1$ type Markov chains[11]. A later book deals at length with the $M/G/1$ type of chain[12]. Multiserver systems with breakdown are studied using these techniques in [13]. The presentation given here is a simplified version of that is given by Latouche and Neuts[10].

13.7   Bibliography

[1]   S. Balsamo and G. Iazeolla. Aggregation and disaggregation in queueing networks. In G. Iazeolla, P.J. Courtois, and A. Hordijk, editors, *Mathematical Computer Performance and Reliability*, pages 95–109. North Holland, Amsterdam, 1984.

[2]   W.-L. Cao and W.J. Stewart. Iterative aggregation/disaggregation techniques for nearly uncoupled Markov chains. *Journal of the ACM*, 32(3):702–719, July 1985.

[3]   P.J. Courtois. Decomposability, instabilities, and saturation in multiprogrammed systems. *Communications of the ACM*, 18(7):371–377, July 1975.

[4]   P.J. Courtois. *Decomposability: Queueing and Computer System Applications*. Academic Press, London, 1977.

[5]   P.J. Courtois. On time and space decomposition of complex structures. *Communications of the ACM*, 28(6):590–603, June 1985.

[6]   R.V. Evans. Geometric distribution in some two-dimensional queueing systems. *Operations Research*, 15(5):830–846, September/October 1967.

[7]   W.K. Grassmann, M.I. Taksar, and D.P. Heyman. Regenerative analysis and steady state distributions for Markov chains. *Operations Research*, 33(5):1107–1116, September/October 1985.

[8]   W.J. Harrod and R.J. Plemmons. Comparison of some direct methods for computing stationary distribution vectors of Markov chains. *SIAM Journal on Computing*, 5(2):453–469, 1984.

[9]   D.P. Heyman. Further comparisons of direct methods for computing stationary distributions of Markov chains. *SIAM Journal on Algebraic and Discrete Methods*, 8(2):226–232, April 1987.

[10] G. Latouche and M.F. Neuts. Efficient algorithmic solutions to exponential tandem queues with blocking. *SIAM Journal on Algebraic and Discrete Methods*, 1(1):93–110, January 1980.

[11] M.F. Neuts. *Matrix-Geometric Solutions in Stochastic Models: An Algorithmic Approach*, volume 2 of *Johns Hopkins Series in the Mathematical Sciences*. The Johns Hopkins University Press, Baltimore, Maryland, 1981.

[12] M.F. Neuts. *Structured Stochastic Matrices of $M/G/1$ Type and Their Applications*. Marcel Dekker, New York, NY, 1989.

[13] M.F. Neuts and D.M. Lucantoni. A Markovian queue with N servers subject to break-downs and repairs. *Management Science*, 25(9):849–861, 1979.

[14] C.C. Paige, G.P.H. Styan, and P.G. Wachter. Computation of the stationary distribution of a Markov chain. *Journal of Statistical Computation and Simulation*, 4(3):173–186, 1975.

[15] G.W. Stewart. Computable error bounds for aggregated Markov chains. *Journal of the ACM*, 30(2):271–285, April 1983.

[16] W.J. Stewart. A new approach to the numerical analysis of Markovian models. In K.M. Chandy and M. Reiser, editors, *Computer Performance*, pages 279–296. North Holland, Amsterdam, 1977.

[17] W.J. Stewart. A comparison of numerical techniques in Markov modelling. *Communications of the ACM*, 21(2):144–152, February 1978.

[18] H. Vantilborgh. Exact aggregation in exponential queueing networks. *Journal of the ACM*, 25(4):620–629, October 1978.

[19] H. Vantilborgh, R.L. Garner, and E.D. Lazowska. Near-complete decomposability of queueing networks with clusters of strongly interacting servers. *Performance Evaluation Review*, 9(2):81–92, 1980.

[20] V.L. Wallace and R.S. Rosenberg. Markovian models and numerical analysis of computer system behaviour. In *Proceedings AFIPS SJCC*, volume 28, pages 141–148, Boston, Massachusetts, 1966.

Chapter 14

Local area networks

In previous chapters we have analysed the performance of various aspects of queueing which may be associated with packet switching communication networks. Local area networks (LANs) have some properties in common with wide area networks, but enough different characteristics to merit different protocol structures. In particular, high bandwidth is available, meaning that efficiency of channel use may not be a prime concern. The size of the network is usually small, so that propagation times are short, and bit error rates are low. Finally, the network is usually owned by the organisation using it, rather than the user having to pay for the bandwidth used. This also reduces the pressure for efficient use of the channel.

Although the usual wide area protocols could be used, and are used in LANs organised as packet switching networks, many LANs do not have the usual multiply connected graph structure, but have different topologies. The two most common are passive buses and active loops. In a bus network, each station is connected to the single cable which forms the LAN. The cable can be split, forming a tree structure, but no loops are allowed. Stations use broadcast protocols in which all stations can receive all transmissions, and if two or more stations transmit at the same time, none of the transmissions succeed. The stations attached to the bus need take no part in network activity until they wish to transmit data. Bus networks usually have no central controller, but have independently operating stations all using the same protocol.

The most popular alternative topology is a ring in which all stations are linked together by a uni-directional channel which forms a single loop. Transmissions propagate round the ring until either they reach their destination, or more commonly, they reach the station at which they started. On ring systems, each station on the ring must be active, even if it does not wish to transmit data, since the data must be forwarded round the ring. There is some resemblance to a store and forward network, except that there is no routing to worry about since all packets take the same route. Usually the stations only store a few bits before forwarding them to the next station, rather than storing whole packets. Ring systems have an element of central control or synchronisation, often provided by a specified monitor station, which ensures that all transmissions succeed.

Other topologies are star networks, in which each station communicates with a central hub, dual cable systems, in which one cable is used for transmission and one for reception, and single cable systems in which each station is attached to the cable at two different points and uses one attachment for transmission and one for reception.

14.1   Broadcast networks

Broadcast local area networks have the property that any packet transmitted by a station can be received successfully by all other stations if and only if only a single station transmits. If two or more stations transmit simultaneously, or if two or more transmissions overlap, then the transmissions are said to collide. In broadcast networks all colliding transmissions will be lost.

The aim of all the protocols that we shall study is to transmit as many packets as possible without collision with transmissions from other stations. An important characteristic is that there should be no central control, all decisions on when, and whether, to transmit a packet being taken on the basis of locally maintained information. In most of our analyses, whether a transmission was successful is assumed to be known. Either a low bandwidth channel can be used to broadcast acknowledgements, or the acknowledgements can form part of the general traffic of the network. In general, a transmitting station is assumed to know within a 'reasonable' time whether it needs to retransmit a packet or can attempt to transmit a new one. Similarly, we ignore the effect of transmission errors on the protocols.

The effectiveness of the protocols can be assessed by calculating the expected number of successful transmissions in an interval. The interval is usually chosen equal to the average transmission time of a packet, so that the expected number of successful transmissions, or *throughput*, usually denoted by S, is always less than 1. The expected number of transmissions attempted in the interval, the *offered traffic*, usually denoted G, can, and often does, exceed 1. The mean delay experienced by packets is also an important performance metric, and the 'throughput-delay' curve can also be studied.

Without specifying any protocol, or performing any analysis, we can make some elementary observations about the relationship of the successful traffic and the offered traffic. When the offered traffic, G, is very small, the successful traffic will be small too. Hence as $G \rightarrow 0$, $S \rightarrow 0$. When the offered traffic is very large, the number of collisions will be large also, so we may expect that as $G \rightarrow \infty$, $S \rightarrow 0$.

14.1.1   ALOHA protocols

The earliest broadcast protocols were developed at the University of Hawaii for use in their radio packet switching network. The protocol that was developed takes its name from that network, and is known as ALOHA. The original protocol, now known as

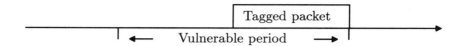

Figure 14.1 Interfering transmissions in pure ALOHA

pure ALOHA, is perhaps the simplest possible technique for scheduling transmissions at distributed locations, requiring no central control at all; whenever a new packet arrives at a station it is transmitted immediately. If no other station's transmission overlaps with the transmission, it will succeed. If the transmission fails, then the packet will be scheduled for retransmission at a random time later. Intuitively, if the traffic is a small fraction of the channel's capacity then a transmission has a good chance of succeeding; on the other hand, if the offered traffic is large, then the chances of some other transmission overlapping are high, and the throughput will be low. Assuming that the offered traffic G, is known, then the successful traffic S, satisfies:

$$S = G \Pr(\text{No collision})$$

In order to make progress towards finding the probability of a transmission being involved in a collision, some assumptions need to be made. It is assumed that all packets are of a constant length, and also that the total offered traffic at all stations, being the combination of the newly arriving packets and the retransmissions of packets which previously collided, is Poisson. It is convenient to express all times in terms of a packet transmission time. This means that the total successful traffic, S, has maximum value 1. Of course, this is in practice unobtainable. Notice that in an environment with no centralised scheduling, such as this one, there is no reason why more than one station should not *attempt* transmission at a time, so the rate of offered traffic, G, can take on any value. In order to calculate the relationship between S and G, consider the transmission of a particular packet. It will fail if any other stations start transmission during the so-called *vulnerable period* for this packet. Figure 14.1 shows the tagged packet starting transmission, and the interval that must be free of other transmissions in order that the transmission should succeed. The vulnerable period starts 1 packet time before the tagged packet's transmission starts: any packet that starts transmission then will still be being transmitted when the tagged packet starts and both transmissions will fail. Similarly, the vulnerable period lasts until the end of transmission of the tagged packet, since any transmission starting during the actual transmission is bound to collide. Hence, the vulnerable period is of length 2, and since transmissions are Poisson at rate G:

$$\Pr(\text{No collision}) = e^{-2G}$$

and:

$$S = Ge^{-2G}$$

The maximum possible S, which is the maximum throughput that the pure ALOHA protocol can support, is calculated by differentiation. The maximum occurs at $G = 0.5$, giving a throughput of $S = 1/(2e)$. Thus with pure ALOHA the maximum possible throughput is approximately 18% of the channel capacity. Even in local area networks, where there is more than adequate bandwidth, this is a rather poor channel utilisation.

The maximum throughput can be increased by reducing the probability of collision. A simple way to achieve that end is to force packet transmissions to be synchronised. In *slotted* ALOHA, all stations keep track of time, and only start transmissions at fixed instants. These instants are separated by intervals equal to one packet transmission time. This means that new packets arriving during a so-called *slot* and retransmissions scheduled for arbitrary times must wait until the start of the next slot before being transmitted. This means that the period during which no new packets must arrive, or retransmissions be scheduled is now the slot before the one in which the tagged packet will be transmitted. This is half the previous interval, so:

$$S = Ge^{-G} \tag{14.1}$$

and the maximum possible throughput is $1/e$, approximately 36% of channel capacity.

This analysis gives the maximum throughput that can be expected with an infinite population of users, or at least a very large population. Finite user populations can be analysed. Consider a slotted ALOHA system, with M stations, and let station i transmit a packet in the next slot with probability g_i. This transmission is independent of transmissions at all other stations, and of the transmission behaviour of this station in other time slots. No distinction is made between new traffic and retransmissions of packets which failed in previous attempts. The probability that station i will successfully transmit a packet is given by:

$$s_i = g_i \cdot \prod_{j \neq i}(1 - g_j)$$

since it must transmit, which has probability g_i, and each of the other stations must remain silent, which occurs with probability $1 - g_j$. The achievable throughput for each station can be calculated, given the rate at which it attempts transmission. The total offered traffic, G, and the total throughput, S, are found by summing over the stations. If the stations are statistically identical, $g_i = g$ and $s_i = s$ for all i, then:

$$s = g(1 - g)^{M-1} \tag{14.2}$$

To find the maximum possible throughput, calculate the roots of the derivative:

$$\frac{\mathrm{d}s}{\mathrm{d}g} = (1-g)^{M-2}(1-Mg)$$

$g = 1$ clearly gives 0 throughput. Maximum throughput is achieved at $g = 1/M$. Hence, G, the total offered traffic, should equal 1 for maximum throughput. Even in asymmetric cases when the offered traffic at different stations is different maximum throughput is achieved at $G = 1$. Notice that s is just the probability that a packet from a particular station, i say, being successfully transmitted will be observed in any slot, s/g is the probability that a transmitted packet is successfully received, and g/s is the expected number of attempted transmissions that a packet will undergo before it is successfully received. These values are often expressed in terms of the network total values, S and G. S is just the probability that a randomly chosen slot will contain a successful packet transmission; G is the expected number of transmissions attempted in a random slot. Expressing equation (14.2) in terms of S and G,

$$S = G(1 - \frac{G}{M})^{M-1}$$

It can be seen that as $M \rightarrow \infty$, this will tend to our infinite population expression, (14.1).

The successful traffic, S, is the rate at which a station can process packets. Notice that for any value of S less than the maximum there are *two* values of G, the rate at which traffic is offered to the network, which give rise to the same value of S. From the discussion of the significance of G/S above, it is clear that the lesser of these G values is preferable.

In order to study further the behaviour of ALOHA, a distinction must be made between new packets and packets which have previously been transmitted and are attempting a retransmission. A station can have at most one packet stored. When the station is idle, a new packet will arrive during a slot and transmission be attempted in the next slot with probability g_N. If this transmission fails, then the station will attempt to transmit the packet in a subsequent slot with probability g_R, independently of the number of transmission attempts that have been made on that packet. The state of the system can be described by the number of stations, b, which have a packet backlogged, awaiting retransmission at the start of each slot. The transition probabilities between the states can easily be calculated. No change of state will occur if no new transmissions are attempted, and none of the backlogged stations attempts transmission. If there are no new transmissions, and two or more backlogged stations attempt transmission, there will be no change of state. A single new transmission which succeeds will not change the state either. The number of backlogged stations decreases by one when a single backlogged station transmits, and there are no new transmissions. Increases in the number of backlogged stations are caused by new transmissions colliding either with other new transmissions, or with retransmission attempts. Writing $p(i, j)$ for the transition probability from the state with i backlogged stations at the start of a slot to j backlogged stations at the start of the next slot, the following formulas apply when $b \neq 0$,

$$
\begin{aligned}
p(b, b) &= (1 - g_N)^{M-b} \cdot (1 - b g_R (1 - g_R)^{b-1}) \\
&\quad + (M - b) g_N (1 - g_N)^{M-b-1} \cdot (1 - g_R)^b \\
p(b, b - 1) &= (1 - g_N)^{M-b} \cdot b g_R (1 - g_R)^{b-1} \\
p(b, b + 1) &= (M - b) g_N (1 - g_N)^{M-b-1} \cdot (1 - (1 - g_R)^b) \\
p(b, b + k) &= \binom{M - b}{k} g_N^k (1 - g_N)^{m-b-k} \\
&\qquad k = 2, \ldots, M - b
\end{aligned}
$$

When $b = 0$, there is no possibility of a transition to state 1, since a single transmission will always succeed, so

$$
\begin{aligned}
p(0, 0) &= (1 - g_N)^M + M g_N (1 - g_N)^{M-1} \\
p(0, k) &= \binom{M}{k} g_N^k (1 - g_N)^{m-k}, \\
&\qquad k = 2, \ldots, M
\end{aligned}
$$

This Markov chain is readily solved numerically, since the transition matrix has only a single entry in each row below the diagonal.

$$
\pi = \pi P
$$

can be expressed as:

$$
\pi_j = \sum_{k=0}^{j+1} \pi_k p(k, j), \quad j = 0, 1, \ldots, M
$$

and hence:

$$
\pi_j = \frac{1}{p(j, j-1)} \left(\pi_{j-1} - \sum_{k=0}^{j-1} \pi_k p(k, j-1) \right), \quad j = 1, 2, \ldots, M
$$

The value of π_0 can then be found from the normalisation condition. The mean number of backlogged stations can be found, and from that the mean delay to a packet is calculated using Little's theorem. Although the mean number of backlogged packets is interesting, it does not reflect the dynamics of the protocol. Examination of the individual state probabilities indicates that the states around the mean have low probability. There are two states which have high probability. Usually one corresponds to the empty state, or a state with only a few backlogged users. The other state with high probability corresponds to a high proportion of the users being backlogged. The system is very likely to be in one of these two states or in one of their close neighbours. It is hardly ever in a state close to the mean.

The system is bi-stable spending long periods of time in the neighbourhood of one of these stable points before it flips into a long period of time in the neighbourhood of the other. The phenomenon can be explained analytically in terms of the drift of the process. The drift of a stochastic process is the expected size of state change it will make, conditioned on the current state. In this case, the expected drift in state i is given by:

$$d(i) = \sum_{j=0}^{M}(j - i)p(i, j)$$

It will be seen that the expected drift in the region of the two stable points is tending to return the system state to one of the stable points.

14.1.2 Carrier sense multiple access protocols

The low value of the maximum channel utilisation possible with an ALOHA protocol can be improved upon by altering the procedure followed. When ALOHA is used, a station will abandon reception to transmit a packet, even if it is receiving a packet addressed to itself. This suggests an immediate improvement in throughput could be made if the station took some account of other transmissions. Instead of transmitting at random, the station which wishes to transmit first of all senses the channel, and if a transmission is detected in progress, the station will refrain from transmission and try again later. Protocols with this characteristic are known as *Carrier-Sense-Multiple-Access* (CSMA). As with ALOHA, slotted and unslotted versions of the protocols exist, depending on whether or not packet transmission starts synchronously at different stations. For analysis purposes, constant length packets are assumed, which have transmission time 1. The other important parameter for analysing CSMA protocols is the maximum time that it takes a signal to propagate from one station to another. This is denoted by a. If the channel is idle, and a station starts transmission at time 0, then another station sensing the channel before a, will detect no transmission in progress and also start transmission causing a collision. These collisions caused by stations starting transmission within a of one another are known as *primary collisions*. Intuitively, if a is short relative to packet transmission time, primary collisions should be rare and throughput should be better than that possible for an ALOHA protocol, because collisions will only occur when two stations start transmission within a of each other.

There are three varieties of CSMA protocol, depending on the action taken when another transmission is sensed. When using a *1-persistent CSMA*, a station sensing a transmission in progress continues to sense the channel, and transmits its packet as soon as the channel is idle. Notice that if two or more stations have all sensed the transmission in progress, then they will all transmit when the channel is idle, and a collision will result. These collisions caused by several stations waiting for another to finish transmission and then all starting transmission simultaneously are known as *secondary collisions*. *Non-persistent* CSMA attempts to avoid these secondary collisions. A station sensing a transmission in progress will defer for a random time, and then try again. This avoids secondary collisions, but still leaves the protocol vulnerable to primary collisions and also tends to increase the delay to packets. $p-persistent$ CSMA takes the end-to-end delay on the channel as a slot time. A station which senses another transmission in progress waits for the channel to be idle in the same way as 1-persistent CSMA. Once the channel is idle, or if no transmission was sensed, then the station will transmit with probability p, and with probability $1-p$, it will defer for one slot, before sensing the channel again. This is an attempt to avoid secondary collisions, as with non-persistent CSMA, but also to avoid the increased delay of non-persistent CSMA. In all three protocols, a transmission which collides with another is retried a random time later. Non-persistent and 1-persistent CSMA both exist in slotted and unslotted versions. $p-persistent$ CSMA has an inherent slotted nature.

Slotted 1-persistent CSMA is analysed as a typical example of the family of protocols. The time slots are chosen to have length a. It is assumed that a packet transmission lasts for an integral number of slots. The total offered traffic, G, made up of newly generated packets and packets rescheduled after previous collisions, is assumed to be Poisson. Hence the probability of i stations starting to sense the channel in any slot is $(aG)^i/i! \cdot e^{-aG}$. In particular, the probability that no stations will start sensing in a slot is e^{-aG}, independently of the number of stations that started sensing in previous slots. A *busy period* is made up of a sequence of *transmission periods*, each consisting of a transmission, or a number of simultaneous transmissions, followed by a single idle slot. Each transmission period is of length $1 + a$, independently of whether or not the transmission succeeded. A transmission period, A, is followed by another transmission period, B, if *any* stations started to sense the channel during A. The transmission in period B will be successful if only one station started sensing during A. If two or more stations started sensing, then all stations sense the idle slot at the end of A, and start transmissions simultaneously; B will then be a wasted transmission period, a secondary collision, and all the stations involved will reschedule their transmissions for a random time in the future. If *no* stations start sensing during A then an idle period starts.

Since traffic is Poisson, the number of scheduled transmissions in each slot is independent of previous slots. There are no transmissions scheduled with probability e^{-aG}. The length of an idle period has a modified geometric distribution with parameter $1 - e^{-aG}$, and hence the mean length of an idle period, I, equals:

$$I = \frac{1}{1 - e^{-aG}} \text{ slots} = \frac{a}{1 - e^{-aG}} \text{ time units}$$

Similarly, the probability of no station starting sensing during a transmission period is $e^{-(1+a)G}$, and the number of transmission periods in a busy period also has a modified geometric distribution, so that the mean length of a busy period, H, is:

$$H = \frac{1 + a}{e^{-(1+a)G}} \text{ time units}$$

In order to calculate the throughput, S, the expected number of *successful* transmission periods in a busy period must be calculated. A transmission period will be successful if and only if a single transmission takes place during that period. The first transmission period of a busy period will be successful provided that only a single station sensed the channel idle in the preceding slot; this has probability ae^{-aG} in a random slot, but in the case of the first transmission period, it is known that at least one station sensed the channel in that slot, hence:

$$\text{Pr(Success in first transmission period)} = \frac{aGe^{-aG}}{1 - e^{-aG}}$$

For the other transmission periods in a busy period, the time interval involved is the previous transmission period rather than the previous slot, hence:

$$\text{Pr(Success in other transmission period)} = \frac{(1 + a)Ge^{-(1+a)G}}{1 - e^{-(1+a)G}}$$

It has already been observed that the expected number of transmission periods in a busy period is geometrically distributed with mean $e^{(1+a)G}$. The mean number of successful transmission periods in a busy period (and hence in a cycle of the system), is thus given by:

$$X = 1 \cdot \frac{aGe^{-aG}}{1 - e^{-aG}} + (e^{(1+a)G} - 1) \cdot \frac{(1 + a)Ge^{-(1+a)G}}{1 - e^{-(1+a)G}}$$

Since the length of each successful transmission is 1, and the mean length of a cycle is $H + I$, the throughput is given by:

$$S = \frac{X}{H + I} = \frac{Ge^{-(1+a)G}(1 + a - e^{-aG})}{(1 + a)(1 - e^{-aG}) + ae^{-(1+a)G}}$$

14.1.3   Carrier sense multiple access with collision detection

The utilisation of CSMA channels can be improved still further by using a carrier sense multiple-access protocol with collision detection. The idea of collision detection is that as well as sensing the channel before it starts transmission, the station continues to sense the channel during its transmission. If any other transmission is sensed, then transmission is aborted immediately, and a jamming signal is broadcast to ensure that all stations have detected the collision. After a station has transmitted for $2a$, all other stations will sense its transmissions in progress, and no collisions can occur. This curtailment of transmissions that are colliding is in contrast to the ALOHA and CSMA protocols which continue to transmit even if a collision is in progress. All these colliding transmissions in CSMA and ALOHA are wasted. As with ordinary CSMA, different CSMA/CD protocols can be defined depending on the action to be taken when the channel is sensed busy.

We shall analyse CSMA/CD assuming a 1-persistent protocol, with retransmissions scheduled for a random time in the future, so that the overall offered traffic consisting of new traffic and retransmissions is Poisson with parameter G. The analysis is similar to that for CSMA, except that transmission periods are not of constant length; if a collision is taking place, all the colliding transmissions will be aborted. Successful transmission periods are of length $1 + a$; unsuccessful transmission periods are of length ja, including the jamming signal transmission. j is one greater than the number of slots for which the jamming signal is transmitted. Since transmissions are scheduled according to a Poisson distribution, the probability that there are *no* transmissions scheduled in a period of length t is e^{-tG}; the probability that there is at least one scheduled is $1 - e^{-tG}$; and the probability that there is exactly one is tGe^{-tG}. As with CSMA, the idle periods are modified geometric distributed with parameter e^{-tG}. Consider the busy periods. They are initiated by the arrivals during a slot, and then continue until there are no arrivals in a transmission period. Denote by $H(t)$ the mean length of a busy period started by the arrivals during an interval of length t. If there is only a single arrival during t, then there will be a single successful transmission period of length $1 + a$, and if there are any arrivals during that transmission period there will be a busy period of mean length $H(1+a)$. On the other hand if there is more than one arrival in t, then a collision period will start with length ja, and it will be followed by a busy period of mean length $H(ja)$, if there were any arrivals.

$$H(t) = \frac{tGe^{-tG}}{1 - e^{-tG}} \left(1 + a + (1 - e^{-(1+a)G})H(1 + a)\right) \tag{14.3}$$

$$+ \left(1 - \frac{tGe^{-tG}}{1 - e^{-tG}}\right) \left(ja + (1 - e^{-jaG})H(ja)\right)$$

Now equation (14.3) can be written with $t = ja$ and with $t = 1 + a$ to give a pair of simultaneous equations in $H(1+a)$ and $H(ja)$. After solution of these equations, the mean busy period is given by $H(a)$. The mean total length of successful transmission periods in a busy period caused by the arrivals in a period of length t is similarly given by $X(t)$, by counting only those transmissions that succeed.

$$
\begin{aligned}
X(t) = & \frac{tGe^{-tG}}{1 - e^{-tG}} \left(1 + (1 - e^{-(1+a)G})X(1 + a)\right) \\
& + \left(1 - \frac{tGe^{-tG}}{1 - e^{-tG}}\right) \left((1 - e^{-jaG})X(ja)\right)
\end{aligned}
$$

As with $H(t)$, two equations in the unknowns $X(1 + a)$ and $X(ja)$ can be solved, and the value of $X(a)$ found by substitution. The throughput is given by:

$$
S = \frac{X(a)}{I + H(a)}
$$

The Ethernet, perhaps the best known CSMA/CD implementation, uses a 1-persistent CSMA/CD protocol with an algorithm known as *binary exponential backoff* to decide on when to transmit packets that were involved in collisions. Each station keeps a count of how many collisions the current packet has been involved in. A packet which has been involved in k collisions is scheduled for retransmission at a time which is chosen to be uniformly distributed between 0 and $2^k \cdot 2a$. (In fact the 'real' Ethernet stops increasing the backoff when k reaches 10, and abandons transmission attempts altogether when k reaches 16.) The binary exponential backoff algorithm is rather tricky to analyse, and most analyses have shown that the retransmission delay after the first collision is the only variable that has a significant effect on the performance of the protocol.

14.2   Ring networks

In ring local area networks, all stations are linked by a uni-directional channel which forms a closed loop. It is possible for different segments of the loop to be of different media, since effectively, the stations are linked by point-to-point connections. The difference from a conventional packet switching network is that whole packets are *not* stored at intermediate nodes. In general, the latency between the first bits of a packet header entering a station on a ring and the same bits leaving that station is very short, usually of the order of a few bit-transmission times.

One of the most important physical parameters of the ring is the *ring delay*, the time it takes a bit to make a circuit of the ring. This is made up of the delay encountered within each station, and the propagation delay through the media which make up the ring. The ring delay corresponds to a number of bits, depending on the ring's transmission rate; these bits can be considered as a (long) shift register. In a ring of physical length L metres, with N stations each of which imposes a further propagation delay of δ, the ring delay is given by:

$$R = \frac{L}{c} + N\delta$$

where c is the propagation speed of electromagnetic waves in the medium. Often this is expressed in terms of a number of bits which are stored in the ring:

$$\beta = \frac{R}{K}$$

where K is the data transmission rate in bits per second.

Within the broad framework of rings, there are three main varieties:

Token Rings   In these rings, a short packet consisting only of a header and small amount of control information circulates while all stations are idle. This *token* is the permission to transmit. A station which has data to send, alters some bits in the header to indicate that the token is in use, and then transmits its data on to the ring. Usually the propagation delay around the ring will be short enough that the header will have returned to the transmitting station before the end of the packet has left the station. When the station has transmitted its data, it regenerates a new free token, and transmits this on the ring.

Slotted Rings   These rings divide the bits travelling round the ring into fixed length slots[†] or mini-packets. The mini-packets all have fields to indicate the source and destination stations, whether the mini-packet is full or empty, and sometimes have error control fields containing parity and/or acknowledgement bits. A station that wishes to transmit has to wait for a mini-packet which is empty to pass, mark it full, and copy in the appropriate addresses and data. The mini-packet is then marked empty either by the receiver of the mini-packet, or by the transmitting station when the mini-packet returns.

Register Insertion Rings   A station wishing to transmit, loads its data into a shift register, and when the end of a packet is detected passing the station, the shift register is switched to become part of the ring. The effective ring delay becomes longer. When the packet returns to the originating station, the shift register is switched out of series with the rest of the ring, and the ring delay is reduced.

Each ring architecture shares the ring's capacity in some way. In a token ring, the capacity is dedicated to the stations in turn; in a slotted ring, the ring's capacity is divided up between the stations that have data to send; and in a buffer insertion ring, the ring's capacity is increased to take account of the stations which are ready to transmit.

[†]The term slot applied to ring local area networks has an entirely different meaning from the slot of CSMA protocols.

Modelling of ring local area networks is particularly complex. The interaction between the different stations and the protocols makes state descriptions rather large. The distributions of waiting times have been analysed under ever more general assumptions, but these are generally beyond the scope of this book. Our analysis will generally keep to the simpler models, where possible using the tools that have been developed in earlier chapters. There will be N stations attached to the ring, each of which has packets arriving at rate λ. As previously, mean delay will be the metric by which our protocols will be measured. Note that the symmetric nature of the traffic considered may give some protocols or architectures advantages over others. It is possible that unbalanced traffic might favour other protocols, but the increased size of the parameter space to be considered has meant that these examples have been omitted. The lengths of packets are arbitrarily distributed, such that the transmission time has distribution function, $B(t)$, Laplace transform, $B^*(s)$, and expectation $\bar{S}$.

14.2.1   Token rings

Token rings are perhaps the most popular architecture for local area networks in use today. An international standard for token ring architecture has been developed by the IEEE 802 committee. This architecture includes various priority levels of traffic. In order to keep the discussion general we will not treat a specific architecture, but deal with a generic token ring.

The generic token ring that we consider has a number of stations. The token circulates, visiting each station in turn until some station has data to transmit. When a station with data to send it must wait until it receives the free token. The token is marked as busy, and the station appends its data to the busy token. In rings with a relatively short ring delay, there will be only a single token; longer rings, whether because of their physical length or because of higher data transmission speeds, may have multiple tokens. In a multiple token ring, when the data has been transmitted, a new token must be inserted immediately. In single token operation, the new token is usually issued after the head of the packet being transmitted has returned to the transmitting station *and* the whole packet has been transmitted. In a single token ring, the effect of the token passing is that each station gets a chance to transmit in turn. In systems where a station may have a queue of packets, there are three forms of service. Exhaustive service allows the station to keep the token until all its packets have been transmitted. Limited service systems restrict the station to send some maximum number of packets, often only one, before passing the token on to another station. Gated service allows the whole queue that was present when the token was acquired to be transmitted, but forces arrivals since the token acquisition to wait for the next appearance of the token.

From the point of view of a station, the token ring appears to function like a server with vacation. The server is absent for a period, and then becomes available to serve the queue at the station. The vacation period is made up of the time that the token takes to propagate round the ring and the busy periods of the other stations on the ring. If the limited service discipline is in use, and the number of packets allowed is one, then the vacation period is a service time for each of the stations which have packets to send, and there is a relaxation period after each service with the same distribution as a vacation. The token will usually take some time to traverse the ring even if the other stations have no packets to send.

The system will be assumed to have M stations, equally spaced along the ring. The token takes a (constant) time, r, to travel form one station to the next. If all stations are idle, the token takes time $R = Mr$ to circulate once round the ring. The stations are statistically identical. At each station, messages arrive at rate λ in a Poisson stream. They have generally distributed lengths, with mean transmission time $\bar{S}$, and second moment $E[S^2]$.

Before analysing the two models, one of exhaustive service, and one of limited service, we make some simple observations about the stability conditions that are needed. Suppose that the system of M queues is stable, that is, a steady state has been reached. The mean time that the token takes to travel once round the ring is made up of the token rotation time, mean R, and the service given to each queue round the ring. Denoting the token's mean cycle time by T, and the mean number of packets transmitted during one cycle by Q, it follows that:

$$T = Q \cdot \bar{S} + R$$

Now if the system is stable, the number of packets which arrive in a cycle must equal the number which are served, so $Q = M\lambda T$, and:

$$T(1 - M\rho) = R$$

where ρ is the offered load at a single station, $\rho = \lambda\bar{S}$. The implication of this relationship is that, in steady state, the token will be found to be free (and circulating between stations) with probability $1 - M\rho$, and conversely, will be found to be in use by some station with probability $M\rho$. Clearly, $\rho < 1/M$ for stability, and:

$$T = \frac{R}{1 - M\rho}$$

The average number of packets transmitted from any particular queue in a cycle is Q/M, and this equals λT.

In a limited service system which serves a maximum of m packets per token visit, there is an additional condition for stability that:

$$\lambda T < m$$

or:

$$\lambda R < m(1 - M\rho)$$

When an exhaustive service system is in use, the system will be stable whatever the value of token rotation time, R. The rationale for this perhaps surprising result is that as the system becomes more and more heavily loaded, less and less of the time is the token free to move from one station to the next.

The tagged job methodology can be used to analyse the exhaustive service token ring system. Consider an arbitrary arriving message; what will its queueing time, w, be? It is made up of the residual service time of the packet currently in service, or the time to reach the next station, if the token is between stations, plus the time to serve all the messages that will be reached before the tagged one, plus the time to traverse the interstation gaps between where the token now is, and the station at which the tagged message has arrived.

$$w = x + n.\bar{S} + g$$

where w is the queueing time, x is the residual service time or residual interstation gap, n the number of messages that will be served before the tagged one[‡], and g the time to traverse the interstation gaps. Taking mean values, in steady state, we have:

$$W = X + N\bar{S} + G \qquad (14.4)$$

X is the mean residual service time, if a service is in progress, and otherwise, the mean residual interstation gap time, hence

$$X = M\rho \frac{M\rho E[S^2]}{2\bar{S}} + (1 - M\rho)\frac{r}{2}$$

Although the system as a whole is in steady state, the state observed by a random observer, or random arrival, will depend on the position of the token—in particular if the token is between stations, the queue at the previous station visited must have all arrived since the token left. Another complication is that not all the messages present in the system will be transmitted before the tagged message. However, careful thought should convince the reader that when the mean number to be served is found, it will still satisfy $N = M\lambda W$[§]. The expected number of interstation gaps that the token needs to cross is simply derived. The best case occurs when the token is being used by the station at which the tagged packet arrived, or when the token is travelling between the previous station and the one at which the tagged arrival occurs. In that case, *no* interstation gaps are crossed. The worst case occurs when the token is in the interstation gap between the tagged station and the next station, or is being used

---

[‡] We have already assumed that the messages are statistically identical, and have mean length $\bar{S}$.

[§] The mean must be considered over all possible stations for the arriving message, as well as all possible possible positions for the token.

by the next station. In that case, the number of gaps to cross is $M - 1$. Associating each interstation gap with the succeeding station, then the total number of gaps that need to be traversed is equiprobably $0, 1, \ldots, M - 1$. Taking the expectation, we find that:

$$G = \frac{M(M - 1)}{2} \cdot \frac{1}{M} \cdot r$$

Equation (14.4) can be rewritten as

$$
\begin{aligned}
(1 - M\rho)W &= X + G \\
&= \frac{M\rho E[S^2]}{2} + (1 - M\rho)\frac{r}{2} + \frac{(M - 1)r}{2} \\
W &= \frac{M\rho E[S^2]}{2(1 - M\rho)} + \frac{(M - M\rho)r}{2(1 - M\rho)} \\
&= \frac{M\rho E[S^2]}{2(1 - M\rho)} + \frac{(1 - \rho)R}{2(1 - M\rho)}
\end{aligned}
$$

Analysis of a gated system can be carried out in exactly the same way. The only quantity that is different is the number of interstation gaps that must be crossed. If the tagged packet arrives when the station currently has the token and is transmitting, then it will have to wait for the next visit of the token to that station before it is transmitted. This will occur with probability ρ, so that G satisfies:

$$G = \frac{M - 1}{2}r + \rho M r$$

and the mean queueing time is:

$$W = \frac{M\rho E[S^2]}{2(1 - M\rho)} + \frac{(1 + \rho)R}{2(1 - M\rho)}$$

When limited service systems are analysed, the number waiting in individual queues needs to be considered. For simplicity, consider the case with only a single packet transmitted at each visit of the token. When a packet arrives at a station queue, its queueing time will be made up of, as before, the residual service time or interstation gap, the service times of the packets that will be served before the tagged packet, and the time spent traversing the ring between stations by the token. The first two quantities have identical derivations to that given above. The number of interstation gaps traversed changes, because the service of a packet results in the token being passed on, independently of whether or not there are more packets to transmit at the station. The expected number of packets waiting at a station is λW, and each of these packets being served causes the token to traverse the ring an extra time. Thus:

$$W = \lambda + M\rho W + G + \lambda W R$$

An exhaustive service version of a limited service system does not make sense. By its nature a limited service system is gated, so the when the appropriate formula for G is included, the queueing time satisfies:

$$W = \frac{M\rho E[S^2]}{2(1 - M\rho - \lambda R)} + \frac{(1 + \rho)R}{2(1 - M\rho - \lambda R)}$$

14.2.2 Slotted rings

Slotted rings treat the whole ring as a long shift register, which is divided up into shorter minipackets or slots. These slots circulate continuously, and each station must be synchronised with the ring in order to know when a new slot starts. Each slot, as well as data fields, contains address fields, and various control bits. The most important control bit is one which indicates whether the slot is in use or not. A station that wishes to transmit, must examine the full/empty bit in each slot until it finds one that is empty. This is then marked as full, the address fields and other control fields initialised, and the data is copied into the slot. When the slot reaches its destination its contents will be copied. Some architectures have the destination station mark the slot as empty. These are known as *destination release* architectures. This requires enough delay in the station for the address and the full/empty bit to both be available. The alternative option is to allow the sender of the slot to mark it as empty. This *source release* is a more common choice, since the latency of the station can be reduced to one or two bits. So long as each station is allowed to have only one transmission on the ring at any time, it will know when the slot in which it transmitted returns just by counting the number of intervening slots. Another design option is whether or not the station which marks a slot as empty can immediately use the same slot for its own transmission and mark it full again. Allowing stations to do this makes a slotted ring a little like a token ring. Most implementations choose not to allow this to happen, which means that a slot always travels empty between the station which emptied it, and the next station which has data to transmit. This *empty slot* principle, guarantees fair sharing of the ring, at the expense of a small amount of wasted bandwidth. A convenient analogy is a circular model railway track, on which as many trucks as possible have been placed and are circulating. Data is sent by waiting for an empty truck to appear, and then filling it. When the truck returns, the data is removed and the truck passed on empty to the next station.

The Cambridge ring is perhaps the best known slotted ring, and it uses the source release rule, and the empty slot principle. Each slot has 38 bits, of which only 16 are data bits. The low level hardware protocol, as well as enforcing the empty slot principle and preventing a station having more than one minipacket on the ring at any time, provides a number of automatic features increasing the response time of the station when the minipacket was not copied. The ring is controlled by a monitor station, which when switched on initialises the ring's slot structure. To do this, it transmits as many slots as will fit on the ring, and then transmits 0 bits as a gap. The first bit in every slot is always 1, so that stations expect a 1 bit every 38 bits. If a 0 bit is read, then the gap has been found, and the station idles until the next 1 bit indicating the start of slots comes around. The monitor station is also responsible for error detection, and clock synchronisation functions, but they do not concern us here.

Modelling the Cambridge ring is made more complex by the addition of software protocols on top of the hardware protocols described in the previous paragraph. The size of the minipacket is such that to allow them to be the sole message size between processes would be almost impossible to program. The hardware provides a solution in terms of a 'Station Select Register' (SSR), in each station which acts as a filter on minipackets. When the SSR is 0, minipackets are accepted from no source, when it is 255, then any minipacket will be accepted, but for any other value of the SSR only minipackets from the station whose address is set in the SSR will be accepted. Minipackets from other stations are ignored (except for setting their response bits). The *packet protocol*, which was originally called the Basic Block protocol, allows larger size packets to be transferred between stations. Stations that wish to transmit repeatedly try to send a packet header to the destination. When the destination accepts a header, it sets its SSR to the address of the source station, making it deaf to all other stations. When the packet has been received, the SSR is set to 255, allowing the station to accept minipackets from any station. Attempting to model this situation in its full generality would be very complex, since a large state space would be needed to describe the exact state of the ring, each station's SSR, the number of minipackets in the queue at each station, etc.

A multilevel model offers a solution. First, a model of the hardware enforced protocol is built, and checked against simulation. This model is then used as the basis of a model of the packet protocol. Consider the hardware protocol first. The empty slot moves round the ring, providing service to all the stations on the ring which wish to transmit. This round-robin scheduling suggests that, provided the mean message length is long enough, the ring can be approximated by a processor sharing server, in which each active station receives an equal share of the ring's bandwidth. The empty slot can be accounted for by an extra fictitious station which is always active. Thus when there are n active stations, each receives $1/(n + 1)$ of the available communications bandwidth. If there are S slots, then the formula usually quoted is that each of n active stations receives a portion $S/(n + S)$ of the rings capacity. To validate this model of the hardware protocol, a simulation of a ring with stations that went through a cycle of 'think'-'transmit P minipackets' was constructed. The actions of the destination stations were ignored, and the mean response time was measured. This was compared with the mean response time from a BCMP model with two nodes; node 1 was an infinite server node, and represented the stations on the ring in think state; the other node represented the ring, and was a processor sharing node, with the rate of service being load dependent. For rings that had only a small number of slots, two or three say, the response time predicted by the model was in close agreement with the simulated response time.

The behaviour of the packet protocol can be modeled by a closed queueing network with $N + 1$ nodes, when the Cambridge ring has N stations. Each station goes through the 'think'-'transmit packet' cycle. There are N jobs in the queueing network, one representing each station. An infinite server node in the queueing network represents the stations that are in think state. The other nodes in the queueing network represent the stations to which packets are addressed. The routing probabilities in the queueing network represent the probability that packets are addressed *to* a particular station. The source of a packet can be 'remembered' by using a multi-class queueing network. The rate of service at the nodes representing the destination stations depends on the number of non-empty queues; that is the number of stations sharing the ring¶. The service rate of each of the nodes representing destination stations is given by $C(k)$ when there are k non-empty queues. This state dependency sets the system outside the separable class of networks which can be solved by the convolution algorithm. The solution can be estimated using the following *fixed point approximation*. Assume that the rate of service at the ring nodes is fixed, C. The network is now separable, and can be solved using any of the standard algorithms. The utilisation, U_i, of each queue is the probability that it is non-empty. Hence, the expected number of non-empty queues is $\sum_{i=1}^{N} U_i$. Taking into account the fact that a transmission *must* be taking place, we find the expected number of non-empty queues is:

$$k^* = \frac{U_1 + U_2 + \cdots + U_n}{1 - p_n}$$

where p_i is the marginal probability that i stations are at node 0, in think state. This allows the rate of service of the ring nodes to be calculated as $C(k^*)$. Now k^* was a function of C, so we have a fixed point relationship that must be satisfied by C if it is to reflect the steady state behaviour of the ring.

$$C = C(k^*) = C(k^*(C))$$

It is possible to show that this relationship has at least one positive real root. Comparison of the results with simulations has shown satisfactory agreement.

¶In fact the total number of jobs attempting to use the ring affects the performance of the whole ring, since minipackets which are ignored are retried, but at a slower rate than ones which are succeeding. We ignore this effect here.

14.2.3   Register insertion rings

Register insertion rings, sometimes known as buffer insertion rings, also work with short packets, although not as short as the slotted ring's minipackets. Each station has several buffers, of sufficient size to take the largest possible packet. Traffic arriving from the ring passes through some recognition circuitry that will decide whether or not the new packet is addressed to this station. Packets that are not addressed to the station are transferred to the insertion buffer. The insertion buffer forms a finite capacity queue. Whenever it is active, data is being transferred from the buffer to the ring, and the queue length in the buffer is being reduced. Packets that are addressed to the station are transferred to a receiving buffer and the queue length in the insertion buffer decreases. Traffic that the station wishes to transmit is loaded into the transmit buffer. If there are sufficient empty queue positions in the insertion buffer, then at the end of the next packet leaving the insertion buffer, the station stops transmitting the insertion buffer, and starts to transmit the contents of the transmit buffer. Any data that arrives from the ring and that needs to be forwarded will accumulate in the insertion buffer. Because the station does not transmit until the insertion buffer has at least enough free capacity to accommodate the packet in the transmit buffer, the insertion buffer cannot overflow. Two different priority schemes are possible. The one just described is *station priority*, because data already on the ring is delayed to allow new data to be inserted. The alternative scheme is *ring priority* in which the traffic being generated by the station is only transmitted to the ring when the insertion buffer is empty. The propagation delay of signals round an otherwise idle register insertion ring is rather larger than in the other two ring architectures because of the necessity to decode addresses before a decision can be made as to whether or not to forward the packet. In slotted ring networks, with source release protocols, only a count of the number of slots seen was needed to know when to empty a slot. In the token ring, at least with single token operation, the new free token can be generated when the head of the packet transmitted by the station returns. Using our model railway analogy, a buffer insertion ring can be thought of as a track with long sidings attached at each station. Data to be sent is loaded into a truck in the siding, and then the points are switched to insert the truck on the 'main line'.

In a packet switching network, the common modelling approach is to assume that packet lengths are independently sampled from a distribution at every link in the network; rather than a packet keeping its length constant throughout its passage through the network. This independence assumption works well when traffic is a mixture of traffic from many sources. In a local area network, the approximation is less valid. Consider for example, a single packet traversing an otherwise idle ring. Its transmission time will be the same at each station it passes through, and so it will not have to queue unless there is other traffic. The independence assumption makes the transmission time differ.

The model of a buffer insertion ring will allow different traffic rates at different stations. In particular, λ_i is the (Poisson) arrival rate of packets at station i. These packets are addressed to station j with probability p_{ij}. Packet transmission times are generally distributed with mean $\bar{S}$ and second moment $E[S^2]$. Consideration of the description above suggests that there should be two queues representing station i. One queue represents the transmit buffer and new packets are added to that queue at rate λ_i. The other queue represents the insertion buffer. We assume that the traffic is such that the overflow of the insertion buffer is impossible. The input to the insertion buffer consists of the output from station $i-1$, less those packets which are addressed to station i. This is assumed to be Poisson at rate γ_i. Using a non-preemptive priority $M/G/1$ model, and defining ρ_i as $\lambda_i \bar{S}$ and η_i as $\gamma_i \bar{S}$, we find that the queueing time of traffic in transit at station i, using the insertion buffer, is given by:

$$D_I^{(r)}(i) = \frac{(\rho_i + \eta_i)E[S^2]}{2(1 - \eta_i)\bar{S}}$$

if ring priority is operated, and:

$$D_I^{(s)}(i) = \frac{(\rho_i + \eta_i)E[S^2]}{2(1 - \eta_i - \rho_i)(1 - \rho_i)\bar{S}}$$

if station priority is used, where the subscript I represents the use of the insertion buffer. Now this traffic in transit is unaffected by the actual behaviour of station $i-1$, whether station or ring priority was applied. If there was *no* new traffic from station i then this traffic would suffer no delay, except the constant time to propagate from station $i-1$ to station i and the latency time at station i required to decode addresses etc. The queueing delay would be zero though. In order that this situation would be correctly dealt with, we introduce a correction term, which involves the subtraction of the standard $M/G/1$ queueing delay:

$$V = \frac{\eta_i E[S^2]}{2(1 - \eta_i)\bar{S}}$$

Hence:

$$V_I^{(r)}(i) = D_I^r(i) - V = \frac{\rho_i E[S^2]}{2(1 - \eta_i)\bar{S}}$$

and for station priority we have:

$$V_I^{(s)}(i) = D_I^{(s)}(i) - V = \frac{\rho_i(1 + \eta_i(1 - \rho_i - \eta_i))E[S^2]}{2(1 - \eta_i - \rho_i)(1 - \rho_i)(1 - \eta_i)\bar{S}}$$

The new traffic which is in the transmit buffer satisfies the ordinary $M/G/1$ non-preemptive priority delay of:

$$V_N^{(r)}(i) = \frac{(\rho_i + \eta_i)E[S^2]}{2(1 - \eta_i - \rho_i)(1 - \eta_i)\bar{S}}$$

when the ring traffic has priority, and:

$$V_N^{(s)}(i) = \frac{(\rho_i + \eta_i)E[S^2]}{2(1 - \rho_i)\bar{S}}$$

when station priority is used. The subscript N indicates the new traffic from station i.

In order to calculate the mean time for a packet to reach its destination, the values of γ_i must be found. Since p_{ij} is the probability that a packet from station i is destined to station j, the probability that a packet from station i will have to pass through the insertion buffer of station k can be found:

$$e_{ik} = \sum_{j \in I(i,k)} p_{ij}$$

where $I(i, k)$ is the set of stations that must be visited by a packet which starts at station i, addressed to station k.

$$I(i, k) = \begin{cases} \{j | k < j \leq i\} & i \geq k \\ \{1, 2, \ldots, i, k + 1, \ldots, M\} & i < k \end{cases}$$

The values of γ_i are then given by:

$$\gamma_i = \sum_{j=1}^{M} \lambda_j e_{ji}$$

The response time for a packet going from station i to station j is made up of its queueing time in the transmit buffer of station i, plus the queueing time in the insertion buffers of all stations between i and j, plus its propagation delay between all these stations, plus its transmission time. Irrespective of whether ring priority or station priority is being used, the formula will be:

$$T_{ij} = V_N(i) + \sum_{k \in M(i,j)} (V_I(k) + \tau_k) + \tau_j + \bar{S}$$

where $M(i, j)$ is the set of nodes that are traversed between i and j, and τ_j is the propagation delay from station $j - 1$ to station j.

$$M(i, j) = \begin{cases} \{k | i < k < j\} & i < j \\ \{1, 2, \ldots, j - 1, i + 1, \ldots, M\} & i \geq j \end{cases}$$

Notice that it is not necessary for all stations to use the same priority system.

14.3   Exercises

1. Analyse the non-persistent CSMA protocol. Assume fixed length packets and a propagation delay of a.

2. A very simple analysis of a CSMA/CD system can be made by assuming that the propagation delay is 0. The state of the system is given by a pair of integers, (i, n), with $i = 0, 1$ representing the number of transmissions in progress, and n representing the number of stations which have found the channel busy and are backing off. Stations which are not currently transmitting attempt to do so at rate λ, and those which are backed off do so at rate ξ. Assume that packets are exponentially distributed with mean length 1, and that all transmission and retransmission attempts are random. Show that these balance equations are satisfied by

$$\pi_{0,n} = \frac{1}{\sigma_n} \pi_{1,n}, \quad n = 0, 1, \ldots, M - 1$$

$$\pi_{1,n} = \left[\binom{M-1}{n} \left(\frac{\lambda}{\xi} \right)^n \prod_{j=1}^{n} \sigma_j \right] \pi_{1,0}, \quad n = 1, 2, \ldots, M - 1.$$

14.4   Further reading

Local area networks has been a very important area of study for a number of years, and the literature is vast. Some early results have been refuted by later more exact analysis, so it is important to read with caution. Hammond and O'Reilly's book[13] is a fairly comprehensive analysis of most types of local network presented as specialisations of a general model.

Broadcast networks started with the ALOHA project at the University of Hawaii, and the first analysis of pure ALOHA was by N. Abramson[1, 2, 3, 4]. The idea of slotting was introduced by L. Roberts[26]. The inherent bistability of ALOHA was analysed by Carleial and Hellman[9]. CSMA was analysed by F. Tobagi and L. Kleinrock[21], and the presentation here is based on that work. F. Tobagi and V.B. Hunt[33] extended the analysis to CSMA/CD. Tasaka's book[31] is a comprehensive account of multiple access protocols. Many other authors have also treated CSMA/CD which is very difficult to analyse exactly, but yields to a number of approximation techniques[22, 29, 10, 15, 19, 23, 11, 30, 32].

Ring local networks date back to the late 1960s when the token ring was developed by Newhall[12]. The slotted ring was suggested by Pierce[25]. The idea of the token ring was taken up by IBM[7] as their local area network offering to customers. Token rings and busses are examples of the general class of polling systems where a number of queues are served one after another. Takagi[27, 28] has given a comprehensive account of the state of the art. Bux has published an excellent survey of analyses of existing systems[6]. The best known slotted ring has been developed at the University of Cambridge. Their architecture has been described in several papers and a book[24, 16, 17]. The analysis suggested here is given in full in [20]. An alternative approach is given by Harrus[14]. Register insertion rings are a slightly later development[18]. The analysis presented here is based on that by W. Bux[8]. He has also made comparative analyses of the three ring architectures[5].

14.5   Bibliography

[1]   N. Abramson. The ALOHA system — another alternative for computer communications. In *Proceedings AFIPS FJCC*, volume 37, pages 281–285, Houston, Texas, 1970.

[2]   N. Abramson. The ALOHA system. In N. Abramson and F.F. Kuo, editors, *Computer Communication Networks*, pages 501–518. Prentice-Hall, Englewood Cliffs, NJ, 1973.

[3]   N. Abramson. The throughput of packet broadcasting channels. *IEEE Transactions on Communications*, COM-25(1):117–128, January 1977.

[4]   N. Abramson. Development of the ALOHANET. *IEEE Transactions on Information Theory*, IT-31(2):119–123, 1985.

[5]   W. Bux. Local area subnetworks: A performance comparison. *IEEE Transactions on Communications*, COM-29(10):1465–1473, October 1981.

[6]   W. Bux. Token ring local-area networks and their performance. *Proceedings IEEE*, 77(2):238–256, February 1989.

[7]   W. Bux, F.H. Closs, K. Kümmerle, H.J. Keller, and H.R. Mueller. Architecture and design of a reliable token-ring network. *IEEE Journal on Selected Areas in Communications*, SAC-1(5):756–765, 1983.

[8]   W. Bux and M. Schlatter. An approximate method for the performance analysis of buffer insertion rings. *IEEE Transactions on Communications*, COM-31(1):50–55, January 1983.

[9]   A.B. Carleial and M.E. Hellman. Bistable behavior of ALOHA-type systems. *IEEE Transactions on Communications*, COM-23(4):401–410, April 1975.

[10] G.L. Choudhury and S.S. Rappaport. Priority access schemes using CSMA-CD. *IEEE Transactions on Communications*, COM-33(7):620–626, July 1985.

[11] E.J. Coyle and B. Liu. Finite population CSMA/CD networks. *IEEE Transactions on Communications*, COM-31(11):1247–1251, November 1983.

[12] W.D. Farmer and E.E. Newhall. An experimental switching system to handle bursty computer traffic. In *Proceedings ACM Symposium Problems Optimization of Data Communication Systems*, pages 1–34, Pine Mountain, Georgia, 1969.

[13] J.L. Hammond and P.J.P. O'Reilly. *Performance Analysis of Local Computer Networks*. Addison-Wesley, Reading, Massachusetts, 1986.

[14] G. Harrus. A model for the basic block protocol of the Cambridge ring. *IEEE Transactions on Software Engineering*, SE-11(1):130–136, 1985.

[15] D.P. Heyman. An analysis of the carrier-sense multiple-access protocol. *Bell System Technical Journal*, 61(8):2023–2051, 1982.

[16] A. Hopper and R.M. Needham. The Cambridge fast ring networking system. *IEEE Transactions on Computers*, C-37(10):1214–1223, October 1988.

[17] A. Hopper and R.C. Williamson. Design and use of an integrated Cambridge ring. *IEEE Journal on Selected Areas in Communications*, SAC-1(5):775–784, 1983.

[18] D.E. Huber, W. Steinlin, and P.J. Wild. SILK: An implementation of a buffer insertion ring. *IEEE Journal on Selected Areas in Communications*, SAC-1(5):766–774, 1983.

[19] Y.C. Jenq. Theoretical analysis of slotted ALOHA, CSMA, and CSMA/CD. In I.F. Blake and H.V. Poor, editors, *Communications and Networks*, pages 325–346. Springer-Verlag, New York, NY, 1986.

[20] P.J.B. King and I. Mitrani. Modelling a slotted ring local area network. *IEEE Transactions on Computers*, C-36(5):554–561, May 1987.

[21] L. Kleinrock and F.A. Tobagi. Packet switching in radio channels: Part I — carrier sense multiple access modes and their throughput-delay characteristics. *IEEE Transactions on Communications*, COM-23(12):1400–1416, December 1975.

[22] S.S. Lam. A carrier sense multiple access protocol for local area networks. *Computer Networks*, 4(1):21–32, 1980.

[23] J. Meditch and C.T.A. Lea. Stability and optimization of the CSMA and CSMA/CD channels. *IEEE Transactions on Communications*, COM-31(6):763–774, June 1983.

[24] R.M. Needham. Systems aspects of the Cambridge ring. In *Proceedings 7th Symposium on Operating System Principles*, pages 82–85, Asilomar, California, 1979.

[25] J.R. Pierce. Network for the block switching of data. *Bell System Technical Journal*, 51(6):1133–1145, 1972.

[26] L.G. Roberts. ALOHA packet system with and without slots and capture. *Computer Communication Review*, 7(1):28–45, 1975.

[27] H. Takagi. *Analysis of Polling Systems*. MIT Press, Cambridge, Massachusetts, 1986.

[28] H. Takagi. Queueing analysis of polling systems. *ACM Computing Surveys*, 20(1):5–28, March 1988.

[29] H. Takagi and L. Kleinrock. Mean packet queueing delay in a buffered two-user CSMA/CD system. *IEEE Transactions on Communications*, COM-33(10):1136–1139, October 1985.

[30] H. Takagi and L. Kleinrock. Throughput analysis for persistent CSMA systems. *IEEE Transactions on Communications*, COM-33(7):627–638, July 1985.

[31] S. Tasaka. *Performance Analysis of Multiple Access Protocols*. MIT Press, Cambridge, Massachusetts, 1986.

[32] F.A. Tobagi. Distribution of packet delay and interdeparture time in slotted ALOHA and carrier sense multiple access. *Journal of the ACM*, 29(4):907–927, October 1982.

[33] F.A. Tobagi and V.B. Hunt. Performance analysis of carrier sense multiple access with collision detection. *Computer Networks*, 4(5):245–259, 1980.

Index